Charles Augustus Young

The Sun

Charles Augustus Young

The Sun

ISBN/EAN: 9783744743013

Printed in Europe, USA, Canada, Australia, Japan

Cover: Foto ©berggeist007 / pixelio.de

More available books at **www.hansebooks.com**

THE INTERNATIONAL SCIENTIFIC SERIES.

THE SUN.

BY

C. A. YOUNG, Ph. D., LL. D.,

PROFESSOR OF ASTRONOMY IN THE COLLEGE OF NEW JERSEY.

WITH NUMEROUS ILLUSTRATIONS.

NEW YORK:
D. APPLETON AND COMPANY,
1, 3, AND 5 BOND STREET.
1881.

PREFACE.

It is my purpose in this little book to present a general view of what is known and believed about the sun, in language and manner as unprofessional as is consistent with precision. I write neither for scientific readers as such, nor, on the other hand, for the masses, but for that large class in the community who, without being themselves engaged in scientific pursuits, yet have sufficient education and intelligence to be interested in scientific subjects when presented in an untechnical manner; who desire, and are perfectly competent, not only to know the results obtained, but to understand the principles and methods on which they depend, without caring to master all the details of the investigation.

I have tried to keep distinct the line between the certain and the conjectural, and to indicate as far as possible the degree of confidence to be placed in data and conclusions.

It is hardly necessary to say that the work has small claims to originality. I have made use of material suited to my purpose from all accessible sources; possibly in some cases (though I hope not) without giving sufficient credit to the original authority. I have been specially indebted to Secchi, Lockyer, Proctor, Ranyard, Vogel, Schellen, and Langley. To the latter, in particular, I am under the greatest obligation for his kindness in preparing for me the notice of his new and important "Bolometric" investigation, which forms the Appendix.

Unforeseen circumstances have caused considerable delay in the printing and publication of the volume. Some remarks, therefore, which were pertinent when put in type last winter, remain so no longer; and certain interesting observations which have been published within the last few months are passed unnoticed.

PRINCETON, *August 1, 1881.*

CONTENTS.

INTRODUCTION.

THE SUN'S RELATION TO LIFE AND ACTIVITY UPON THE EARTH.

                                                                                                PAGE

Brief Statement of the Principal Facts relating to the Sun, and of the Accepted Views as to its Constitution 11

CHAPTER I.

DISTANCE AND DIMENSIONS OF THE SUN.

Importance of the Problem.—Definition of Parallax.—Aristarchus's Determination of the Parallax.—Different Available Methods.—Observations of Mars.—Transits of Venus.—Observations of Contacts and Photographic Work.—Determination of Solar Parallax by means of the Velocity of Light; by Lunar and Planetary Perturbations.—Illustrations of the Immensity of the Sun's Distance.—Diameter of the Sun.—The Sun's Mass and Density 20

CHAPTER II.

METHODS AND APPARATUS FOR STUDYING THE SURFACE OF THE SUN.

Projection of Solar Image upon a Screen.—Carrington's Method of determining the Position of Objects on Sun's Surface.—Solar Photography.—Photoheliographs.—Cornu's Methods.—Telescope with Silvered Object-Glass.—Herschel's Solar Eyepiece.—The Polarizing Eyepiece 50

CHAPTER III.

THE SPECTROSCOPE AND THE SOLAR SPECTRUM.

The Spectrum and Fraunhofer's Lines.—The Prismatic Spectroscope; Description of Various Forms and Explanation of its Operation.—The Diffraction Spectroscope.—Analyzing and Integrating Spectroscopes.—The Telespectroscope and its Adjustment.—Explanation of Lines in the Spectrum.—Kirchhoff's Researches and Laws.—The Sun's Absorbing Atmosphere and Reversing Layer.—Elements present in the Sun.—Lockyer's Researches and Hypothesis.—Basic Lines.—Dr. H. Draper's Investigations as to the Presence of Oxygen in the Sun.—Schuster's Observations.—Effect of Motion upon Wave-Length of Rays and Spectroscopic Determinations of Motion in Line of Sight 66

CHAPTER IV.

SUN-SPOTS AND THE SOLAR SURFACE.

Granulation of Solar Surface.—Views of Langley, Nasmith, Secchi, and others.—Faculæ.—Nature of the Photosphere.—Janssen's Photographs of Solar Surface—the *Resau Photospherique.*—Discovery of Sun-spots.—General Appearance and Structure of a Spot.—Its Formation and Disappearance.—Duration of Sun-spots.—Remarkable Phenomena observed by Carrington and Hodgson.—Observations of Peters.—Dimensions of Spots.—Proof that Spots are Cavities.—Sun-spot Spectrum.—"Veiled Spots."—Rotation of Sun.—Equatorial Acceleration.—Explanations of the Acceleration.—Position of Sun's Axis and Secchi's Table for its Position Angle at Different Times of the Year.—Proper Motions of Spots.—Distribution of Spots . . . 102

CHAPTER V.

PERIODICITY OF SUN-SPOTS; THEIR EFFECTS UPON THE EARTH, AND THEORIES AS TO THEIR CAUSE AND NATURE.

Observations of Schwabe.—Wolf's Numbers.—Proposed Explanations of Periodicity.—Connection between Sun-Spots and Terrestrial Magnetism.—Remarkable Solar Disturbances and Magnetic Storms.—Effect of Sun-Spots on Temperature.—Sun-Spots, Cyclones, and Rainfall.—Researches of Symons and Meldrum.—Sun-Spots and Commercial Crises.—Galileo's Theory of Spots.—Herschel's Theory.—Secchi's First Theory.—Zöllner's.—Faye's.—Secchi's Later Opinions.—Other Theories 144

CHAPTER VI.

THE CHROMOSPHERE AND THE PROMINENCES.

Early Observations of Chromosphere and Prominences.—The Eclipses of 1842, 1851, and 1860.—The Eclipse of 1868.—Discovery of Janssen and Lockyer.—Arrangement of Spectroscope for Observations upon Chromosphere.—Spectrum of Chromosphere.—Lines always present.—Lines often reversed.—Motion Forms.—Double Reversal of Lines.—Distribution of Prominences.—Magnitude.—Classification of Prominences as quiescent, and eruptive or metallic.—Isolated Clouds.—Violence of Motion.—Observations of August 5, 1872.—Theories as to the Formation and Causes of the Prominences 179

CHAPTER VII.

THE CORONA.

General Appearance of the Phenomenon.—Various Representations.—Eclipses of 1857, 1860, 1867, 1868, 1869, 1871, and 1878.—Proof that the Corona is mainly a Solar Phenomenon.—Brightness of the Corona.—Connection with Sun-Spot Period.—Spectrum of the Corona.—Application of the Analyzing and Integrating Spectroscopes.—Polarization.—Evidence of the Slitless Spectroscope as to the Constitution of the Corona.—Changes and Motions in the Corona.—Its Form and Constitution, and Theories as to its Nature and Origin 213

CHAPTER VIII.

THE SUN'S LIGHT AND HEAT.

Sunlight expressed in Candle-Power.—Method of Measurement.—Brightness of the Sun's Surface.—Langley's Experiment.—

Diminution of Brightness at Edge of the Sun's Disk.—Hastings's View as to Nature of the Absorbing Envelope.—Total Amount of Absorption by Sun's Atmosphere.—Thermal, Luminous, and Actinic Rays: their Fundamental Identity and Differences.—Measurement of the Sun's Radiation.—Herschel's Method.—Expressions for the Amount of Sun's Heat.—Pouillet's Pyrheliometer.—Crova's.—Violle's Actinometer.—Absorption of Heat by Earth's Atmosphere; by the Sun's.—Question as to Differences of Temperature on Different Portions of Sun's Disk.—Question as to Variation of Sun's Radiation with Sun-Spot Period.—The Sun's Temperature—Actual—Effective.—Views of Secchi, Ericsson, Pouillet, Vicaire, and Rosetti.—Evidence from the Burning-Glass.—Langley's Experiment with the Bessemer "Converter."—Permanency of Solar Heat for last Two Thousand Years.—Meteoric Theory of Sun's Heat.—Helmholtz's Contraction Theory.—Possible Past and Future Duration of the Sun's Supply of Heat 240

CHAPTER IX.

SUMMARY OF FACTS, AND DISCUSSION OF THE CONSTITUTION OF THE SUN.

Table of Numerical Data.—Constitution of Sun's Nucleus.—Peculiar Properties of Gases under High Temperature and Pressure.—Characteristic Differences between a Liquid and a Gas.—Constitution of the Photosphere and Higher Regions of the Sun's Atmosphere.—Professor Hastings's Theory.—Pending Problems of Solar Physics 278

APPENDIX.

Professor Langley's Account of his Bolometric Observations and Certain Conclusions derivable from them 298

THE SUN.

INTRODUCTION.

THE SUN'S RELATION TO LIFE AND ACTIVITY UPON THE EARTH.

Brief Statement of the Principal Facts relating to the Sun, and of the Accepted Views as to its Constitution.

It is true that from the highest point of view the sun is only one of a multitude—a single star among millions—thousands of which, most likely, exceed him in brightness, magnitude, and power. He is only a private in the host of heaven.

But he alone, among the countless myriads, is near enough to affect terrestrial affairs in any sensible degree; and his influence upon them is such that it is hard to find the word to name it; it is more than mere control and dominance. He does not, like the moon, simply modify and determine certain more or less important activities upon the surface of the earth, but he is almost absolutely, in a material sense, the prime mover of the whole. To him we can trace directly nearly all the energy involved in all phenomena, mechanical, chemical, or vital. Cut off his rays for even a single month, and the earth would die; all life upon its surface would cease.

There always has been a more or less distinct recognition of this fact. The first man's experience of the first sunset ever witnessed by human eyes must have made it tremendously obvious, when he saw the sun descend below the horizon, and the darkness close in upon the earth, and felt the chill of night, and fell asleep not knowing of a sunrise to come—unless, perhaps, some divine revelation took pity on the hopeless terror he must otherwise have suffered, or unless he may have been, like a little child, slow to notice and unable to comprehend what would frighten a more intelligent being.

But while the material supremacy of the sun has always been recognized by thoughtful minds, and has even been made the foundation of religious systems, as with the Persians, it has been reserved for more modern times, and to our own century, to show clearly just how, in what sense, and how far the sunbeams are the life of the earth, and the sun himself the symbol and vicegerent of the Deity. The two doctrines of the correlation of forces and the conservation of energy, having once been distinctly apprehended and formulated, it has been comparatively easy to confirm them by experiment and observation, and then to trace, one by one, to their solar origin, the different classes of energy which present themselves in terrestrial phenomena—to show, for instance, how the power of waterfalls is only a transformation of the sun's heat; and that the same thing is true, a little more remotely but just as certainly, of the power of steam, of electricity, and even of animals. The idea is now so familiar that it is hardly necessary to dwell upon it, and yet, for some of our readers at least, it may be worth while to examine it a little more closely.

Whenever work is done, it is by the undoing of some previous work. When a clock moves, it is the unwinding of a spring or the falling of a weight which keeps it going, and some one must have wound it up to begin with. If the water of a river falls year after year over a cataract, and is intercepted to drive our mill-wheels, the river continues to run because some power is continually raising and returning to the hill-tops the water which has flowed into the sea—a process precisely equivalent to the daily rewinding of the clock. If the powder in a rifle explodes and drives out the bullet, its explosive energy depends upon the fact that some power has placed the component molecules in such relations that, when the trigger is pulled, and the exciting spark has, so to speak, cut the bonds which hold them apart, they rush together just as suspended weights would fall if freed. Before the same substance, which once was a charge of gunpowder, but now is dust and gas, can again do the same work, the products of the explosion must by some power be decomposed, and the atoms replaced in the same relations as before the firing of the gun; and this process is mechanically analogous to the lifting of fallen weights and placing them upon elevated shelves, or hanging them from hooks, ready to drop again when the occasion may require.

Precisely the same thing is true of the heat produced by the combustion of ordinary fuel: it is due to the collapse of molecules, for the most part of oxygen on one side, and carbon and hydrogen on the other, which have been separated and built up into structures by the action of some laboring power.

The same can be said of animal power, for all investigation goes to show that in a mechanical sense the body of an animal is only a very ingenious and effective machine, by means of which the living inhabitant which controls it can utilize the energy derived from the food taken into the stomach. The body, regarded as a mechanism, is only a food-engine in which the stomach and lungs stand for the furnace and boiler of a steam-engine, the nervous system for the valve-gear, and the muscles for the cylinder. *How* the personality within, which wills and acts, is put into relation with this valve-gear, so as to determine the movements of the body it resides in, is the inscrutable mystery of life; the facts in the case, however, being no less facts because inexplicable.

And now, when we come to inquire for the source of the energy which lifts the water from the sea to the mountain-top, which decomposes the carbonic acid of the atmosphere, and plant-foods of the soil, and builds up the hydrocarbons and other fuels of animal and vegetable tissue, we find it always mainly in the solar rays. I say mainly because, of course, the light and heat of the stars, the impact of meteors, and the probable slow contraction of the earth, are all real sources of energy, and contribute their quota. But, as compared with the energy derived from the sun, their total amount is probably something like the ratio of starlight to sunlight;* so small that it is quite clear, as we said

---

* About forty years ago, Pouillet came to a conclusion entirely inconsistent with the statement of the text. From his actinometric observations, he deduced a temperature of — 224° F. (— 142 C.) for the "temperature of space," which is 236° (131 C.) above the absolute zero. To maintain this temperature of — 224°, he calculated that the stars and space in general must furnish to the earth about 85 per cent. as much heat as the sun supplies. His calculations, however, rest upon assumptions as to the laws of cooling and radiation which are not at present received as accurate, and he fails to take proper account of the influence of water-vapor in the air—an influence, the magnitude of which was first brought out more than twenty years later by the researches of Tyndall and Magnus. It is now generally admitted, therefore, that his result can not be accepted.

INTRODUCTION.

before, that a month's deprivation of the solar rays would involve the utter destruction of all activity upon the earth.

It is natural, therefore, that modern science should make much of the sun, and that the study of solar phenomena and relations should be pursued with the greatest interest. For the last thirty years this has been especially the case: Schwabe's discovery of the periodicity of the sun-spots in 1851; the development of spectroscopic analysis between 1854 and 1870; the eclipse observations since 1860; the researches of Carrington, Huggins, De La Rue, Lockyer, Janssen, Secchi, Vogel, Langley, and others; the establishment of the solar observatories at Potsdam and Meudon—these are all evidences of the ardor with which astronomers have devoted themselves to the problems of solar science, and of their rich rewards.

It may be well, before entering upon the more extended discussion of our subject, to summarize here a few of the more important and obvious facts relating to the sun, with a brief statement of the views at present generally held in regard to its constitution.

To the few unaided eyes which are able to bear its brilliance without flinching, the sun presents the appearance of a round, white disk, a little more than half a degree in diameter—i. e., a row of seven hundred suns, side by side, would just about fill up the circle of the horizon. Usually, without a telescope, the surface appears simply uniform, except that there is a slight darkening at the edge, and that once in a while black spots are seen upon the disk. There is nothing in the sun's appearance to indicate his real distance, and, until that is known, of course no conclusion can be arrived at as to his true dimensions; but the heat of his rays is obvious, and, long before the days of telescopes and thermometers, led to the conclusion that he is nothing more or less than an enormous ball of fire.

If we watch him from day to day through the year, beginning about the 21st of March, we shall find that at noon he daily rises higher in the heavens, until about the 22d of June; at this time he ascends to the same height each noon for several successive days, and then slides slowly south, passing on September 22d the elevation he had at starting, and keeping on until, on December 21st, he attains his farthest southing; thence he returns, till he reaches the place of beginning, and night and day again are equal.

If, at the same time, one has noticed the stars each night, he will find the constellations to have shifted with the months, in such a way that it is clear that the sun has been traveling eastward among them through the sky, as well as swinging north and south; moving, in fact, yearly around the heavens in a path which is a great circle of the globe, inclined some $23\frac{1}{2}°$ to the equator, and called the ecliptic, because it is only when the moon is near this line at new or full that eclipses happen.

There is nothing in this motion which of itself can inform us whether its cause is a real movement of the sun around the earth, or of the earth around the sun. At present, of course, every one knows that the earth is really the moving body. A careful watching shows that her path is not quite circular, or, at least, that the sun is not exactly in the center, since it is one hundred and eighty-four days through the summer from the vernal to the autumnal equinox, and only one hundred and eighty-one from the autumnal to the vernal.

This much was known to the ancients, and the one further fact that the sun's distance is many times greater than that of the moon; it is all that could possibly be learned without the use of the telescope and instruments of precision.

Modern astronomy has gone much further. We now know that the sun's average distance from the earth is about 93,000,000 miles, and consequently that his diameter is about 865,000 miles. The sun has been weighed against the earth and found to contain a quantity of matter nearly 330,000 times as great, and comparing this with his enormous bulk, it appears that his mean density is only about one fourth that of the earth, or one and a quarter times that of water —in other words, the *mass* of the sun is about one fourth greater than that of a globe of water of the same size.

The visible surface of the sun has been named the *photosphere*, and by watching the spots, which occasionally appear upon it, we have ascertained that it revolves upon its axis once in about twenty-five and a quarter days. At times of total eclipse, when the moon hides from us the body of the sun, we are enabled to see certain outlying phenomena at other times invisible. We find close around the luminous surface a rose-colored stratum of gaseous matter to which Frankland and Lockyer some years ago assigned the name of *chromosphere*. Here and there great masses of this chromospheric matter rise high above the general level like clouds of flames, and are then known as *prominences* or *protuberances*.

Outside of the chromosphere is the mysterious *corona*, an irregular halo of faint, pearly light, composed for the most part of radial filaments and streamers, which extend outward from the sun to an enormous distance; often more than a million of miles.

The spectroscope informs us that, in great part at least, the elements, which exist in the lower regions of the solar atmosphere in the state of vapor, are the same we are familiar with upon the earth; while it shows the chromosphere and prominences to consist mainly of hydrogen, and makes it possible to observe them even when the sun is not hidden by the moon. The secret of the corona it fails to unlock as yet, though it informs us of the presence in it of an unknown gas of inconceivable tenuity.

The *pyrheliometer* and *actinometer* measure for us the outflow of solar heat, and show us that the blaze is at least seven or eight times as intense as that of any furnace known to art. The quantity of heat emitted is enough to melt a shell of ice ten inches thick over the whole surface of the sun every second of time: this is equivalent to the consumption of a layer of the best anthracite coal nearly four inches thick every single second.

Combining the facts just stated, astronomers are for the most part agreed upon the following conclusions as to the constitution of the sun:

1. The central portion is probably for the most part a mass of intensely heated gases.

2. The photosphere is a shell of luminous clouds, formed by the cooling and condensation of the condensible vapors at the surface, where exposed to the cold of outer space.

3. The chromosphere is composed mainly of uncondensible gases (conspicuously hydrogen) left behind by the formation of the photospheric clouds, and bearing something the same relation to them that the oxygen

and nitrogen of our own atmosphere do to our own clouds.

4. The corona as yet has received no explanation which commands universal assent. It is certainly truly solar to some extent, and very possibly may be also to some extent meteoric

CHAPTER I.

DISTANCE AND DIMENSIONS OF THE SUN.

Importance of the Problem.—Definition of Parallax.—Aristarchus's Determination of the Parallax.—Different Available Methods.—Observations of Mars.—Transits of Venus.—Observations of Contacts and Photographic Work.—Determination of Solar Parallax by means of the Velocity of Light; by Lunar and Planetary Perturbations.—Illustrations of the Immensity of the Sun's Distance.—Diameter of the Sun.—The Sun's Mass and Density.

The problem of finding the distance of the sun is one of the most important and difficult presented by astronomy. Its importance lies in this, that this distance—the radius of the earth's orbit—is the base-line by means of which we measure every other celestial distance, excepting only that of the moon; so that error in this base propagates itself in all directions through all space, affecting with a corresponding proportion of falsehood every measured line—the distance of every star, the radius of every orbit, the diameter of every planet.

Our estimates of the masses of the heavenly bodies also depend upon a knowledge of the sun's distance from the earth. The quantity of matter in a star or planet is determined by calculations whose fundamental data include the distance between the investigated body and some other body whose motion is controlled or modified by it; and this distance generally enters into the computation by its cube, so that any error in it involves a more than threefold error in the resulting mass. An uncertainty of one per cent. in the sun's distance implies an uncertainty of more than three per cent. in every celestial mass and every cosmical force.

Error in this fundamental element propagates itself in time also, as well as in space and mass. That is to say, our calculations of the mutual effects of the planets upon each other's motions depend upon an accurate knowledge of their masses and distances. By these calculations, were our data perfect, we could predict for all futurity, or reproduce for any given epoch of the past, the configurations of the planets and the conditions of their orbits, and many interesting problems in geology and natural history seem to require for their solution just such determinations of the form and position of the earth's orbit in by-gone ages.

Now, the slightest inaccuracy in the data, though hardly affecting the result for epochs near the present, leads to error which accumulates with extreme rapidity in the lapse of time; so that even the present uncertainty of the sun's distance, small as it is, renders precarious all conclusions from such computations when the period is extended more than a few hundred thousand years. If, for instance, we should find as the result of calculation with the received data, that two millions of years ago the eccentricity of the earth's orbit was at a maximum, and the perihelion so placed that the sun was nearest during the northern winter (a condition of affairs which it is thought would produce a glacial epoch in the southern hemisphere), it might easily happen that our results would be exactly contrary to the truth, and that the state of affairs indicated did not occur within half a million years of the specified date—and all because in our calculation the sun's distance, or solar parallax by which it is measured, was assumed half of one per cent. too great or too small. In fact, this solar parallax enters into almost every kind of astronomical computations, from those which deal with stellar systems and the constitution of the universe, to those which have for their object nothing higher than the prediction of the moon's place as a means of finding the longitude at sea.

Of course, it hardly need be said that its determination is the first step to any knowledge of the dimensions and constitution of the sun itself.

This parallax of the sun is simply *the angular semi-diameter of the earth as seen from the sun;* or, it may be defined in another way as the angle between the direction of the sun ideally observed from the center of the earth, and its actual direction as seen from a station where it is just rising above the horizon.

We know with great accuracy the dimensions of the earth. Its mean equatorial radius, according to the latest and most reliable determination (agreeing, however, very closely with previous ones), is 3962·720 English miles [6377·323 kilometres], and the error can hardly amount to more than $\frac{1}{100000}$ of the whole—perhaps, 200 feet one way or the other. Accordingly, if we know how large the earth looks from any point, or, to speak technically, if we know the parallax of the point, its distance can at once be found by a very easy calculation: it equals simply [206,265 * × the radius of the earth] ÷ [the parallax in seconds of arc].

---

* This number 206,265 is the length of the radius of a circle expressed in seconds of its circumference. A ball one foot in actual diameter would have an apparent diameter of one second at a distance of 206,265 feet, or a little more than 39 miles. If its apparent diameter were 10″, its distance would, of course, be only $\frac{1}{10}$ as great.

DISTANCE AND DIMENSIONS OF THE SUN. 23

Now, in the case of the sun it is very difficult to find the parallax with sufficient precision on account of its smallness—it is less than 9″, almost certainly between 8·75″ and 8·85″. But this tenth of a second of doubtfulness is more than $\frac{1}{100}$ of the whole, although it is no more than the angle subtended by a single hair at a distance of nearly 800 feet. If we call the parallax 8·80″, which is probably very near the truth, the distance of the sun will come out 92,885,000 miles, while a variation of $\frac{1}{20}$ of a second either way will change it about half a million of miles.

When a surveyor has to find the distance of an inaccessible object, he lays off a convenient base-line, and from its extremities observes the directions of the object, considering himself very unfortunate if he can not get a base whose length is at least $\frac{1}{10}$ of the distance to be measured. But the whole diameter of the earth is less than $\frac{1}{11000}$ of the distance of the sun, and the astronomer is in the predicament of a surveyor who, having to measure the distance of an object ten miles off, finds himself restricted to a base of less than five feet; and herein lies the difficulty of the problem.

Of course, it would be hopeless to attempt this problem by direct observations, such as answer perfectly in the case of the moon, whose distance is only thirty times the earth's diameter. In her case, observations taken from stations widely separated in latitude, like Berlin and the Cape of Good Hope, or Washington and Santiago, determine her parallax and distance with very satisfactory precision; but if observations of the same accuracy could be made upon the sun (which is not the case, since its heat disturbs the adjustments of an instrument), they would only show the parallax to be somewhere between 8" and 10"; its distance between 126,-000,000 and 82,000,000 miles.

Astronomers, therefore, have been driven to employ indirect methods based on various principles: some on observations of the nearer planets, some on calculations founded upon the irregularities—the so-called perturbations—of lunar and planetary movements, and some upon observations of the velocity of light. Indeed, before the Christian era, Aristarchus of Samos had devised a method so ingenious and pretty in theory that it really deserved success, and would have attained it were the necessary observations susceptible of sufficient accuracy. Hipparchus also devised another founded on observations of lunar eclipses, which also failed for much the same reasons as the plan of Aristarchus.

The idea of Aristarchus was to observe carefully the number of hours between new moon and the first quarter, and also between the quarter and the full. The first interval should be shorter than the second, and the difference would determine how many times the distance of the sun from the earth exceeds that of the moon, as will be clear from the accompanying figure. The moon reaches its quarter, or appears as a half-moon,

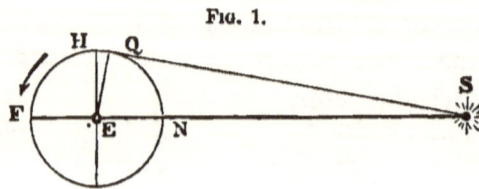

Fig. 1.

when it arrives at the point Q, where the lines drawn from it to the sun and earth are perpendicular to each other. Since the angle H E Q = E S Q, it will follow that H Q is the same fraction of H E as Q E is of E S; so that, if H Q can be found, we shall at once have the ratio of Q E and E S. Aristarchus thought he had ascertained that the first quarter of the month (from N to Q) was about 12 hours shorter than the second, from which he computed the sun to be about 19 times as distant as the moon. The difficulty lies in the impossibility of determining the precise moment when the disk of the moon is exactly bisected, and depends partly upon the fact that the lunar surface is very rough and broken, and partly upon the fact that the sun's diameter is nearly twice that of the orbit of the moon, instead of being a mere point as in the figure. The consequence is, that there is no sharp boundary between light and darkness; the *terminator*, as it is called, is both irregular and ill-defined. The real difference between the first and second quarters is not quite 36 minutes, so that the sun's distance is about 400 times the moon's.

The different methods upon which our present knowledge of the sun's distance depends may be classified as follows:

1. Observations upon the planet Mars near opposition, in two distinct ways:
   (a) Observations of the planet's declination made from stations widely separated in latitude.
   (b) Observations from a single station of the planet's right ascension when near the eastern and western horizons—known as Flamsteed's or Bond's method.
2. Observations of Venus at or near inferior conjunctions:
   (a) Observations of her distance from small stars measured at stations widely different in latitude.
   (b) Observations of the transits of the planet: 1. By noting the *duration* of the transit at widely-separated stations; 2. By noting the true Greenwich time of contact of the planet with the sun's limb; 3. By measuring the distance of the planet from the sun's limb with suitable micrometric apparatus; 4. By photographing the transit, and subsequently measuring the pictures.

26   THE SUN.

3. By observing the oppositions of the nearer asteroids in the same manner as those of Mars.
4. By means of the so-called parallactic inequality of the moon.
5. By means of the monthly equation of the sun's motion.
6. By means of the perturbations of the planets, which furnish us the means of computing the ratios between the *masses* of the planets and the sun, and consequently their distances—known as Leverrier's method.
7. By measuring the velocity of light, and combining the result
   (*a*) with equation of light between the earth and sun; or
   (*b*) with the constant of aberration.

Our scope and limits do not, of course, require or allow any exhaustive discussion of these different methods and their results, but some of them will repay a few moments' consideration:

The first three methods are all based upon the same general idea, that of finding the actual distance of one of the nearer planets by observing its displacement in the sky as seen from remote points on the earth. The *relative* distances of the planets are easily found in several different ways,* and are known with very great

* One method of determining the relative distances of a planet and the sun from each other and from the earth is the following, known since the days of Hipparchus: First, observe the date when the planet comes

FIG. 2.

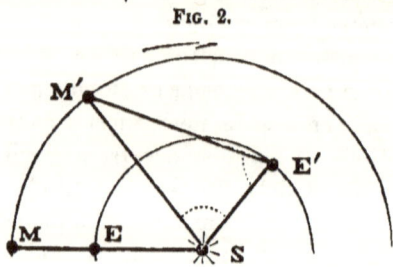

to its opposition—i. e., when sun, earth, and planet are in line, as in the figure, where the planet and earth are represented by M and E. Next, after a known number of days, say one hundred, when the planet has ad-

accuracy—the possible error hardly reaching the ten-thousandth in even the most unfavorable cases. In other words, we are able to draw for any moment an exceedingly accurate map of the solar system—the only question being as to the scale. Of course, the determi-

Fig. 3.

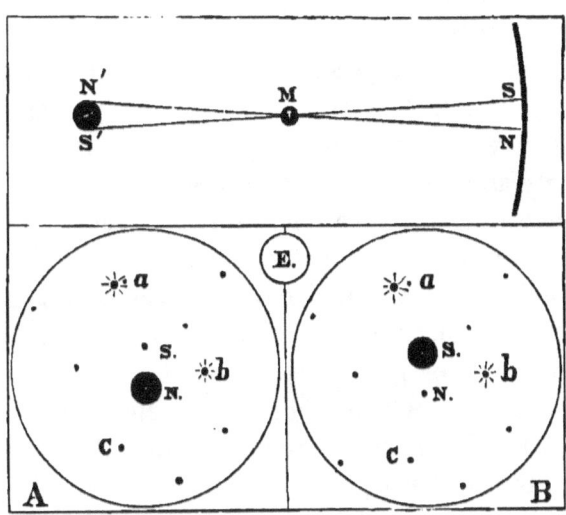

nation of any line in the map will fix this scale; and for this purpose one line is as good as another, so that the measurement of the distance from the earth to the planet Mars, for instance, will settle all the dimensions of the system.

vanced to M and the earth to E', observe the planet's elongation from the sun, i. e., the angle M' E' S. Now, since we know the periodic times of both the earth and planet, we shall know both the angle M S M' moved over by the planet in one hundred days, and also E S E' described in the same time by the earth. The difference is M S E, often called the synodic angle. We have, therefore, in the triangle M' S E', the angle at E' measured, and the angle M' S E' known as stated above; this of course gives the third angle at M', and hence by the ordinary processes of trigonometry we can find the relative values of its three sides.

Fig. 3 illustrates the method of observation. Suppose two observers, situated one near the north pole of the earth, the other near the south. Looking at the planet, the northern observer will see it at N (in the upper figure), while the other will see it at S, farther north in the sky. If the northern observer sees it as at A (in the lower part of the figure), the southern will at the same time see it as at B; and, by measuring carefully at each station the apparent distance of the planet from several of the little stars (a, b, c) which appear in the field of view, the amount of the displacement can be accurately ascertained. The figure is drawn to scale. The circle E being taken to represent the size of the earth as seen from Mars when nearest us, the black disk represents the apparent size of the planet on the same scale, and the distance between the points N and S, in either figure A or B, represents, on the same scale also, the displacement which would be produced in the planet's position by a transference of the observer from Washington to Santiago, or *vice versa*.

The first modern attempt to determine the sun's parallax was made by this method in 1670, when the French Academy of Sciences sent Richer to Cayenne to observe the opposition of Mars, while Cassini (who proposed the expedition), Roemer, and Picard observed it from different stations in France. When the results came to be compared, however, it was found that the planet's displacement was imperceptible by their existing means of observation: from this they inferred that the planet's parallax could not exceed half a minute of arc, and that the sun's could not be more than 10″.

In 1752 Lacaille at the Cape of Good Hope made similar observations, and their comparison with corresponding observations in Europe showed that instruments had so far improved as to make the displacement quite sensible. He fixed the sun's parallax at $10''$, corresponding to a distance of about 82,000,000 miles.

In more recent times the method has been frequently applied, and with results on the whole satisfactory. In 1849–'52 Lieutenant Gilliss was sent by the United States Government to Santiago, in Chili, to observe both Mars and Venus in connection with northern observatories. In 1862 a still more extended campaign was organized, in which a great number of observatories in both hemispheres participated. Professor Newcomb's careful reduction of the work puts the resulting parallax at $8.855''$. The method can be used to the best advantage, of course, when at the time of opposition the planet is near its perihelion and the earth near its aphelion; these favorable oppositions occur about once in fifteen years, and the one which occurred in September, 1877, was so exceptionally advantageous that great pains were taken to secure its careful and general observation.

The expedition of Mr. Gill to the Island of Ascension deserves special notice, since his methods of observation were to some extent different from any before employed, and more refined; and his results seem to be entitled to a very great weight as compared with others.

His instrument was a so-called *heliometer*, loaned by Lord Lindsay for the expedition. It consists essentially of a telescope, having its object-glass divided into two semicircular pieces, which can be made to slide by each other. Each half of the lens makes its own image of the object under examination, so that, by properly setting the lenses, the images of two neighboring stars can be brought to coincide; and, if we know the displacement of the lenses, which can be measured by an accurate scale, the angular distance of the stars can be determined with a precision unattainable by any other known method. The instrument is delicate and complicated, but, in the hands of an observer who understands it, is thoroughly reliable.

Mr. Gill's method is a modification of Flamsteed's. His observations consisted in measurements of the apparent distance between the planet and the stars lying near its path, the work being kept up each night during nearly the whole time of the planet's visibility above the horizon.

About three hundred and fifty sets of such measures were obtained, and all the principal observatories co-operated in the work by determining with their utmost precision the places of the comparison stars.

The final reductions have not yet appeared (May, 1880), but in 1879 he announced $8·783'' \pm 0·015''$ as a preliminary and very approximate value of the solar parallax, which can not be changed to any appreciable extent by the small corrections still remaining to be applied to the star-places.

So far as can be judged from the work thus far published, this determination must be conceded the precedence over all others in respect to its probable freedom from constant and systematic errors, and from theoretical difficulties.

In observations of this sort upon Mars or the asteroids, the position and displacement of the planet, as seen from different stations, are determined by comparing it with neighboring stars. When Venus, however, is nearest us, she can be observed only by day, so that in her case star comparisons are as a general thing out of the question. But occasionally at her inferior conjunction she passes directly across the disk of the sun, and her parallactic displacement from different stations can then be determined by making any such observations as will enable the computer to ascertain accurately her apparent distance and direction from the sun's center at some given moment. Gregory in 1663 first pointed out the utility of such observations for ascertaining the parallax, but it was not until some fifteen years later that the attention of astronomers was secured to the subject by Halley, who discussed the matter thoroughly, and showed how the problem might be solved with accuracy by observations such as were entirely practicable even by the instruments and with the knowledge then at command. In 1761 and 1769 two transits occurred which were observed in all accessible quarters of the globe by expeditions sent out by the different governments. From different sets of these observations variously combined by different computers, values of the solar parallax were obtained ranging all the way from $7.5''$ to $9.2''$. A general discussion of all the materials afforded by the two transits was first made by Encke in 1822, and he obtained, as the most probable result, the value $8.5776''$, which from that time for more than thirty years was accepted by all astronomers as the best attainable approximation to the truth. In 1854 Hansen, in publishing some of his results respecting the motion of the moon, announced that Encke's value of the solar parallax could not be reconciled with his investigations; within the next six or seven years several independent researches by other astronomers confirmed his conclusions, and the most recent recomputations by Powalky, Stone, Faye, and others, show that the errors of observation were so considerable in 1769 that nothing more can be fairly deduced from that transit than that the solar parallax is probably somewhere between 8·7″ and 8·9″.

The method of observation then used consisted simply in noting the moment when the limb of the planet came in contact with that of the sun—an observation which is attended with much more difficulty and uncertainty than would at first be supposed. The difficulties depend in part upon the imperfections of optical instruments and the human eye, partly upon the essential nature of light, leading to what is known as diffraction, and partly upon the action of the planet's atmosphere. The two first-named causes produce what is

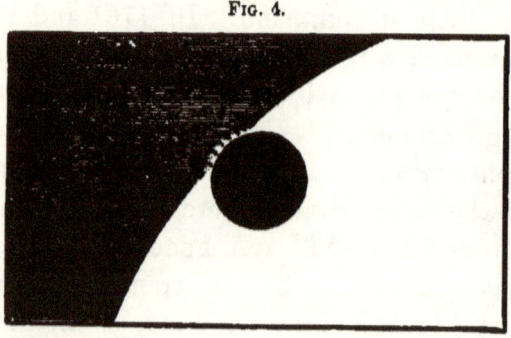

Fig. 4.

called irradiation, and operate to make the apparent diameter of the planet, as seen on the solar disk, smaller than it really is—smaller, too, by an amount which varies with the size of the telescope, the perfection of its lenses, and the tint and brightness of the sun's image. The edge of the planet's image is also rendered slightly hazy and indistinct.

The planet's atmosphere also causes its disk to be surrounded by a narrow ring of light, which becomes visible long before the planet touches the sun, and at the moment of internal contact produces an appearance of which the accompanying figure is intended to give an idea, though on an exaggerated scale. The planet moves so slowly as to occupy more than twenty minutes in crossing the sun's limb; so that, even if the planet's edge were perfectly sharp and definite, and the sun's limb undistorted, it would be very difficult to determine the precise second at which contact occurs; but as things are, observers, with precisely similar telescopes, and side by side, often differ from each other five or six seconds; and where the telescopes are not similar, the differences and uncertainties are much greater. The extent of the difficulty can be judged of by the simple fact that, from the whole mass of contact observations, obtained in 1874 by the different British parties which observed the transit, three different values of the solar parallax have been deduced by different computers, viz., the official value $8.76''$ by Airy, $8.81''$ by Tupman, and $8.88''$ by Stone. These differences depend mainly upon the different interpretations given to the description of phenomena noted by the observers in the field. Very little seems to have been gained in this respect since 1769. Astronomers, therefore, at present are pretty much agreed that such observations can be of little value in removing the remaining uncertainty of the parallax, and are disposed to put more reliance upon the micrometric and photographic methods, which are free from these peculiar difficulties, though of course beset with others; which, however, it is hoped will prove less formidable.

The micrometric method requires the use of a heliometer, an instrument common only in Germany, and requiring much skill and practice in its use in order to obtain with it accurate measures. At the late transit a single English party, two or three of the Russian parties, and all five of the German, were equipped with these instruments, and at some of the stations extensive series of measures were made. None of the results, however, have appeared as yet, so that it is impossible to say how greatly, if at all, this method will have the advantage in precision over the contact observations.

The Americans and French placed their main reliance upon the photographic method, while the English and Germans also provided for its use to a certain extent. The great advantage of this method is that it makes it possible to perform the necessary measurements, upon whose accuracy everything depends, at leisure after the transit, without hurry, and with all possible precautions. The field-work consists merely in obtaining as many and as good pictures as possible. A principal objection to the method lies in the difficulty of obtaining good pictures, i. e., pictures free from distortion, and so distinct and sharp as to bear high magnifying power in the microscopic apparatus used for their measurement. The most serious difficulty, however, is involved in the accurate determination of the scale of the picture; that is, of the number of seconds of arc corresponding to a linear inch upon the plate. Besides this, we must know the exact Greenwich time at which each picture is taken, and it is also extremely desirable that the *orientation* of the picture should be accurately determined; that is, the north and south, east and west points of the solar image on the finished plate. There has been a good deal of anxiety lest the image, however accurate and sharp when first produced, should alter in course of time through the contraction of the collodion film on the glass plate, but the experiments of Rutherfurd, Huggins, and Paschen seem to show that this danger is imaginary; that if a plate is properly prepared the collodion film never creeps at all, but remains firmly attached to the glass. It requires but a very trifling amount of distortion or inaccuracy of the image to render it useless. The uncertainty in our present knowledge of the sun's parallax is so small that it would only involve an error of about one quarter of a second in the calculated position of Venus on the sun's disk as seen from any station at any given time during the transit, and this would be about $\frac{1}{2000}$ of an inch on a four-inch picture of the sun. Unless, then, the picture is so distinct and free from distortion that the relative positions of Venus and the sun's center can be determined from it within $\frac{1}{2000}$ of an inch, it is worthless as a means of correcting the received determination of the parallax.

But it is to be noted that any mere enlargement or diminution of the diameter of sun or planet will do no harm, provided it is alike all around the circumference of the disk, since the measurement is not from the edge of Venus to the edge of the sun, but between their *centers*. Photographic determinations of *contact*, on the contrary (such as Janssen and some of the English parties attempted by a peculiar and complicated apparatus), are affected with all the uncertainties of the old-fashioned observations of the eye alone, and with others in addition; so that, astronomically considered, they are entirely worthless, although interesting from a chemical and physical point of view.

Two essentially different lines of proceeding were adopted, at the last transit, in the photographic observations. The English and Germans attached a camera to the eye-end of an ordinary telescope, which was pointed directly at the sun; the image formed at the focus of the telescope was enlarged to the proper size by a combination of lenses in the camera; and a small plate of glass ruled with squares was placed at the focus of the telescope and photographed with the sun's image, furnishing a set of reference-lines, which give the means of detecting and allowing for any distortion caused by the enlarging lenses.

The Americans and French, on the other hand, preferred to make the picture of full size, without the intervention of any enlarging lens: as this requires an object-glass with a focal length of thirty or forty feet, which could not be easily pointed at the sun, a plan proposed first by M. Laussedat, but also independently by our own Professor Winlock, was adopted. The telescope is placed horizontal, and the rays are reflected into the object-glass by a plane mirror suitably mounted. The French used mirrors of silvered glass, and took their pictures (about two and a half inches in diameter) by the old daguerreotype process on silvered plates of copper, in order to avoid the risk of collodion-contraction. With the silvered mirror the time of exposure is so short that no clock-work is required. The Americans used *unsilvered* mirrors, in order to avoid any distorting action of the sun's rays upon the form of the mirror. This, of course, made the light feebler, and the time of exposure longer, so that a clock-work movement of the mirror was needed to keep the image from changing its place on the plate during the exposure, which, however, never exceeded half a second. The American pictures were taken by the ordinary wet process on glass, and were about four inches in diameter. Just in front of the sensitive plate, at a distance of about one eighth of an inch, was placed a reticle, or a plate of glass ruled in squares, and between this and the collodion-plate hung a fine silver wire suspending a

DISTANCE AND DIMENSIONS OF THE SUN. 37

plumb-bob. Thus the finished negative was marked into squares, and also bore the image of the plumb-line, which, of course, indicated precisely the direction of the vertical. The Americans also placed the photographic telescope exactly in line with a meridian instrument, and so determined, with the extremest precision, the direction in which it was pointed. Knowing this, and the time at which any picture was taken, it becomes

FIG. 5.

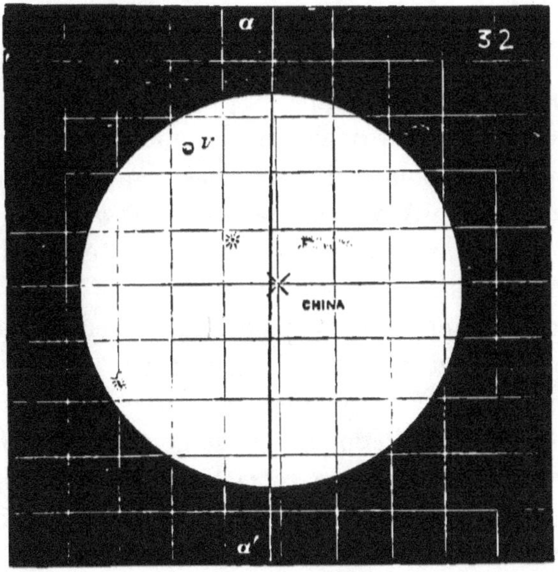

possible, with the help of the plumb-line image, to determine precisely the orientation of the picture—an advantage possessed by the American pictures alone, and making their value nearly twice as great as otherwise it would have been.

The above figure is a representation of one of the American photographs reduced about one half. V is the image of Venus, which on the actual plate is about one seventh of an inch in diameter; $a\ a'$ is the image of the plumb-line. The center of the reticle is marked by the little cross, and the word "China," written on the reticle-plate with a diamond—and, of course, copied on the photograph—indicates that it is one of the Peking pictures. Its number in the series is given in the right-hand upper corner. About 90 such pictures were obtained at Peking during the transit, and about 350 at all the eight American stations, the work being much interfered with by unfavorable weather at most of them. If we add those obtained by the French, Germans, and English, the total number available reaches nearly 1,200, according to the best estimates.

After the pictures are made and safely brought home, they have next to be measured—i. e., the distance (and in the American pictures the *direction* also) between the center of Venus and the center of the sun must be determined in each picture. This is an exceedingly delicate and tedious operation, rendered more difficult by the fact that the image of the sun is never truly circular, but, even supposing the instrument to be perfect in all its adjustments, is somewhat distorted by the effect of atmospheric refraction; so that the true position of the sun's center with reference to the squares of the reticle is determined only by an intricate calculation from measurements made with a microscopic apparatus on a great number of points suitably chosen on the circumference of the image. The final result of the measurement comes out something in this form: Peking. No. 32. Time, $14^h\ 08\ 20\cdot2^s$ (Greenwich mean time); Venus north of sun's center, $735\cdot32''$; east of center, $441\cdot63''$; distance from center of sun, $857\cdot75''$. (The numbers given are only imaginary.) These measurements and calculations are understood to have been for some time completed on the American pictures, but for some reason they have not been published, and there has been a delay in the subsequent work of combining the results and deducing the most probable value of the parallax. The delay is unfortunate, as, in view of the approaching transit of 1882, it is becoming important to know whether the method has any real value. There is evidently a growing apprehension that no photographic process can be relied on.

The English photographs proved of little value. They were measured by two different persons, and from the measurements of one (Mr. Burton) a parallax of 8·25″ was deduced, while from those of the other (Captain Tupman) the result was 8·08″. One of the principal difficulties evidently lay in the uncertainty of the scale-value, which was only deduced from the diameters of the sun and planet; a difficulty from which the American photographs are free, as their scale-value was independently and precisely determined by an accurate measurement of the distance between the lens and the plate on which the picture was taken.

The English photographic results are the only ones which have yet appeared.

One of the best methods of determining the solar parallax is based upon the careful observation of the motions of the moon. The first suspicion as to the correctness of the then received distance of the sun was raised in 1854 by Hansen's announcement that the moon's parallactic inequality led to a smaller value than that deduced from the transit of Venus—a conclusion corroborated by Leverrier four years later, from the so-called lunar equation of the sun's motion. It seems at first sight strange, but it is true, as Laplace long since pointed out, that the skillful astronomer, by merely watching the movements of our satellite, and without leaving his observatory, can obtain the solution of problems which, attacked by other methods, require tedious and expensive expeditions to remote corners of the earth. Our scope and object do not require us to enter into detail respecting this lunar method of finding the sun's parallax; it must suffice to say that the disturbing action of the sun makes the interval from new moon to the first quarter about eight minutes longer than that from the quarter to full; and this difference depends upon the *ratio between the diameter of the moon's orbit and the distance of the sun* in such a manner that, if the inequality is accurately observed, the ratio can be calculated. Since we know the distance of the moon, this will give that of the sun. The results obtained in this way, according to the most recent investigations, appear to fix the solar parallax between $8·83''$ and $8·92''$.

But the method by which ultimately we shall obtain the most accurate determination of the dimensions of our system is that proposed by Leverrier, depending upon the secular perturbations produced by the earth upon her neighboring planets; especially in causing the motions of their nodes and perihelia. These motions are very slow, but *continuous;* and hence, as time goes on, they will become known with ever-increasing accuracy. *If they were known with absolute precision, they would enable us to compute, with absolute precision also, the ratio between the masses of the sun and earth*, and from this ratio we can calculate * the distance of the sun by either of two or three different methods.

* One method of proceeding is as follows: Let M be the mass of the sun, and *m* that of the earth; let R be the distance of the sun from the earth, and *r* that of the moon; finally, let T be the number of days in a

As matters stand at present, the majority of astronomers would probably consider that these secular perturbations are not yet known with an exactness sufficient to render this method superior to the others that have been named—perhaps as yet not even their rival. Leverrier, on the other hand, himself put such confidence in it that he declined to sanction or coöperate in the operations for observing the recent transit of Venus, considering all labor and expense in that direction as merely so much waste.

But, however the case may be now, there is no question that as time goes on, and our knowledge of the planetary motions becomes more minutely precise, this method will become continually and cumulatively more exact, until finally, and not many centuries hence, it will supersede all the others that have been described. The parallax of the sun, determined by Leverrier in this method, in 1872, comes out $8 \cdot 86''$.

The last of the methods mentioned in the synopsis given on pages 25 and 26 is interesting as an example of the manner in which the sciences are mutually connected and dependent. Before the experiments of Fizeau in 1849, and of Foucault a few years later, our knowledge of the velocity of light depended on our knowledge of

sidereal year, and t the number in a sidereal month. Then, by elementary astronomy—

$$M : m = \frac{R^3}{T^2} : \frac{r^3}{t^2}; \text{ whence } R^3 = r^3 \left(\frac{T^2}{t^2}\right) \left(\frac{M}{m}\right);$$

or, in words, *the cube of the sun's distance equals the cube of the moon's distance, multiplied by the square of the number of sidereal months in a year, and by the ratio between the masses of the sun and earth.* It is to be noted, however, that T and t are the periods of the earth and moon, as they would be if wholly undisturbed in their motions, and hence differ slightly from the periods actually observed—the differences are small, but somewhat troublesome to calculate with precision.

the dimensions of the earth's orbit. It had been found by astronomical observations upon the eclipses of Jupiter's satellites that light occupied a little more than sixteen minutes in crossing the orbit of the earth, or about eight minutes in coming from the sun; and hence, supposing the sun's distance to be 95,600,000 miles, as was long believed, the velocity of light must be about 192,000 miles per second. Thus optics was indebted to astronomy for this fundamental element. But when Foucault in 1862 announced that, according to his unquestionably accurate experiments, the velocity of light could not be much more than 186,000 miles per second, the obligation was returned, and the suspicions as to the received value of the sun's parallax, which had been raised by the lunar researches of Hansen and Leverrier, were changed into certainty. The experimental determinations of the velocity of light by Cornu in 1873–'74 fix the solar parallax between 8·78″ and 8·85″, according as we use Struve's "constant of aberration" or Delambre's value of the "equation of light," which is the name given to the time required for light to traverse the interval between the sun and the earth.

Very recently, in 1878–'79, Master A. A. Michelson, of the United States Navy, has made a new and exceedingly accurate measurement of the velocity of light by a modification of Foucault's method. His result is 299,920 kilometres, or 186,360 miles per second, agreeing with that of Cornu within about forty miles per second, and almost certainly not in error by an amount so large as this forty miles.

Mr. D. P. Todd has since then published a careful discussion of the sun's parallax as deduced from the velocity of light, and finds 8·808″; the limits of error, in his opinion, lying between 8·78″ and 8·83″. There

DISTANCE AND DIMENSIONS OF THE SUN. 43

are, however, some possible objections to the method, depending on the uncertainty as to the velocity of the motion of the solar system in space, and the possible effect of this motion upon the propagation of light. The necessary correction can not be large, but its exact amount can not be determined by any method yet known to science.

Collecting all the evidence at present attainable, it would seem that the solar parallax can not differ much from $8·80''$, though it may be as much as $0·02''$ greater or smaller; this would correspond, as has already been said, to a distance of 92,885,000 miles, with a probable error of about one quarter of one per cent., or 225,000 miles.*

But, though the distance can easily be stated in figures, it is not possible to give any real idea of a space so enormous; it is quite beyond our power of conception. If one were to try to walk such a distance, supposing that he could walk 4 miles an hour, and keep it up for 10 hours every day, it would take $68\frac{1}{2}$ years to

---

* The oscillations of scientific opinion as to the value of this constant have been very curious. Early in the century Laplace, in the "Mécanique Céleste," adopted the value $8·81''$ given by the first discussion of the transits of Venus in 1761–'69; but other astronomers, Delambre, for instance, proposed a smaller value. Encke, as has been said before, made a new and thorough discussion of these transits in 1822–'24, and deduced the value $8·58''$, which held the ground for nearly forty years. About 1860 the researches of Hansen, Leverrier, and Stone were thought to have established a value exceeding $8·90''$, and the "British Nautical Almanac" still uses $8·95''$. In 1865 Newcomb published a careful investigation, based upon all the data then known, and deduced the value $8·848''$. Leverrier, in 1872, found $8·86''$ from the planetary perturbations. The "American Ephemeris" and the Berlin "Jahrbuch" use Newcomb's value, and the French "Connaissance de Temps" employs Leverrier's. It appears, however, perfectly certain, from the work of the last few years, that the figures ($8·80''$) given in the text are much nearer to the truth.

make a single million of miles, and more than 6,300 years to traverse the whole.

If some celestial railway could be imagined, the journey to the sun, even if our trains ran 60 miles an hour, day and night and without a stop, would require over 175 years. Sensation, even, would not travel so far in a human lifetime. To borrow the curious illustration of Professor Mendenhall, if we could imagine an infant with an arm long enough to enable him to touch the sun and burn himself, he would die of old age before the pain could reach him, since, according to the experiments of Helmholtz and others, a nervous shock is communicated only at the rate of about 100 feet per second, or 1,637 miles a day, and would need more than 150 years to make the journey. Sound would do it in about 14 years if it could be transmitted through celestial space, and a cannon-ball in about 9, if it were to move uniformly with the same speed as when it left the muzzle of the gun. If the earth could be suddenly stopped in her orbit, and allowed to fall unobstructed toward the sun under the accelerating influence of his attraction, she would reach the center in about four months. I have said if she could be stopped, but such is the compass of her orbit that, to make its circuit in a year, she has to move nearly 19 miles a second, or more than fifty times faster than the swiftest rifle-ball; and in moving 20 miles her path deviates from perfect straightness by less than one eighth of an inch. And yet, over all the circumference of this tremendous orbit, the sun exercises his dominion, and every pulsation of his surface receives its response from the subject earth.

By observing the slight changes in the sun's apparent diameter, we find that its distance varies somewhat at different times of the year, about 3,000,000 miles in all; and minute investigation shows that the earth's orbit is almost an exact ellipse, whose nearest point to the sun, or *perihelion*, is passed by the earth about the 1st of January, at which time she is 91,385,000 miles distant.

The distance of the sun being once known, its dimensions are easily ascertained—at least, within certain narrow limits of accuracy. The angular semi-diameter of the sun when at the mean distance is almost exactly 962″, the uncertainty not exceeding $\frac{1}{2000}$ of the whole. The result of twelve years' observations at Greenwich (1836 to 1847) gives 961·82″, and other determinations oscillate around the value first mentioned, which is that adopted in the "American Nautical Almanac." Taking the distance as 92,885,000 miles, this makes the sun's diameter 866,400; and the probable error of this quantity, depending as it does *both* on the error of the measured diameter and of the distance, is some 4,000 or 5,000 miles; in other words, the chances are strong that the actual diameter is between 860,000 and 870,000 miles.

Measurements made by the same person, however, and with the same instrument, but at different times, sometimes differ enough to raise a suspicion that the diameter is slightly variable, which would be nothing surprising considering the nature of the solar surface.

There is no sensible difference between the equatorial and polar diameters, the rotation of the sun on its axis not being sufficiently rapid to make the polar compression (which must, of course, necessarily result from the rotation) marked enough to be perceived by our present means of observation.

It is not easy to obtain any real conception of the

vastness of this enormous sphere. Its diameter is 109·5 times that of the earth, and its circumference proportional; so that the traveler who could make the circuit of the world in 80 days would need nearly 24 years for his journey around the sun. Since the surfaces of spheres vary as the squares, and bulks as the cubes, of their diameters, it follows that the sun's surface is nearly 12,000 times, and its volume, or bulk, more than 1,300,000 times, greater than that of the earth. If the earth be represented by one of the little three-inch globes common in school apparatus, the sun on the same scale will be more than 27 feet in diameter, and its distance nearly 3,000 feet. Imagine the sun to be hollowed out and the earth placed in the center of the shell thus formed, it would be like a sky to us, and the moon would have scope for all her motions far within the inclosing surface; indeed, since she is only 240,000 miles away, while the sun's radius is more than 430,000, there would be room for a second satellite 190,000 miles beyond her.

The *mass* of the sun, or quantity of matter contained in it, can also be computed when we know its distance, and comes out nearly 330,000 times as great as the earth. The calculation may be made either by means of the proportion given in the note to page 40, or by comparing the attracting force of the sun upon the earth, as indicated by the curvature of her orbit (about 0·119 inch per second), with the distance a body at the surface of the earth falls in the same time under the action of gravity, a quantity which has been determined with great accuracy by experiments with the pendulum. Of course, the fact that the sun produces its effect upon the earth at a distance of 92,885,000 miles, while a falling body at the level of the sea is only about 4,000

DISTANCE AND DIMENSIONS OF THE SUN. 47

miles from the center of the attraction which produces its motion, must also enter into the reckoning.*

This mass, if we express it in pounds or tons, is too enormous to be conceived: it is 2 octillions of tons— that is, 2 with 27 ciphers annexed; it is nearly 750 times as great as the combined masses of all the planets and satellites of the solar system—and Jupiter alone is more than 300 times as massive as the earth. The sun's attractive power is such that it dominates all surrounding space, even to the fixed stars, so that a body at the distance of our nearest stellar neighbor, a Centauri, which is more than 200,000 times remoter than the sun, could free itself from the solar attraction only by darting away with a velocity of more than 300 feet per second, or over 200 miles an hour; unless animated by a greater velocity than this, it would move around the sun in a closed orbit—an ellipse of some shape, or a circle—with a period of revolution which, in the smallest possible orbit, would be about 31,600,000 years, and if the orbit were circular, would be nearly 90,000,000. We say it would revolve thus—that is, of course, unless

---

* The calculation of the sun's mass, from the data given, proceeds as follows: Let M = the sun's mass, and m that of the earth; R = the distance from the earth to the sun, and r the mean radius of the earth; T, the length of the sidereal year, reduced to seconds; and $\frac{1}{2} g$ the distance a body falls in a second at the earth's surface. Now, the distance the earth falls toward the sun in a second, or the curvature of her orbit in a second, is equal to $\dfrac{2\pi^2 R}{T^2}$ (about 0·119 inch). Hence, by the law of gravitation, $\frac{1}{2} g : \dfrac{2\pi^2 R}{T^2} = \dfrac{m}{r^2} : \dfrac{M}{R^2}$, whence, $M = m \left(\dfrac{4\pi^2 R^3}{T^2 r^2 g} \right)$.

In this formula make $\pi = 3\cdot14159$; R, 92,885,000 miles; $T = 31{,}558{,}149\cdot3$ seconds; $r = 3{,}958\cdot2$ miles: and $g = 0\cdot0061035$ mile (16·113 feet), and we shall get the result given in the next, viz., $M = 330{,}000\,m$ (nearly).

intercepted or diverted from its course by the influence of some other sun, as it probably would be. And we may notice here that in many cases certainly, and in most cases probably, the stars are flying through space at a far swifter rate, with velocities of many miles per second.

As for the attraction between the sun and earth, it amounts to thirty-six hundred quadrillions of tons: in figures, 36 followed by seventeen ciphers. On this point we borrow an impressive illustration from a recent calculation by Mr. C. B. Warring. We may imagine gravitation to cease, and to be replaced by a material bond of some sort, holding the earth to the sun and keeping her in her orbit. If now we suppose this connection to consist of a web of steel wires, each as large as the heaviest telegraph-wires used (No. 4), then to replace the sun's attraction these wires would have to cover the whole sunward hemisphere of our globe about as thickly as blades of grass upon a lawn. It would require *nine* to each square inch.

If we calculate the force of gravity at the sun's surface, which is easily done by dividing its mass, 330,000, by the square of $109\frac{1}{2}$ (the number of times the sun's diameter exceeds the earth's), we find it to be $27\frac{1}{2}$ times as great as on the earth; a man who on the earth would weigh 150 pounds, would there weigh nearly two tons; and, even if the footing were good, would be unable to stir. A body which at the earth falls a little more than 16 feet in a second would there fall 443. A pendulum which here swings once a second would there oscillate more than five times as rapidly, like the balance-wheel of a watch—quivering rather than swinging.

Since the sun's volume is 1,300,000 times that of the earth, while its mass is only 330,000 times as great, it follows at once that the sun's *average density* (found by dividing the mass by the volume) *is only about one quarter that of the earth*. This is a fact of the utmost importance in its bearing upon the constitution of this body. As we shall see hereafter, we know that certain heavy metals, with which we are familiar on the earth, enter largely into the composition of the sun, so that, if the principal portion of the solar mass were either solid or liquid, its mean density ought to be at least as great as the earth's; especially since the enormous force of solar gravity would tend most powerfully to compress the materials. The low density can only be accounted for on the supposition, which seems fairly to accord also with all other facts, that the sun is mainly a ball of gas, or vapor, powerfully condensed, of course, in the central portion by the superincumbent weight, but prevented from liquefaction by an exceedingly high temperature. And, on the other hand, it could be safely predicted on physical principles that so huge a ball of fiery vapor, exposed to the cold of space, would present precisely such phenomena as we find by observation of the solar surface and surroundings.

CHAPTER II.

METHODS AND APPARATUS FOR STUDYING THE SURFACE OF THE SUN.

Projection of Solar Image upon a Screen.—Carrington's Method of determining the Position of Objects on Sun's Surface.—Solar Photography.—Photoheliographs.—Cornu's Methods.—Telescope with Silvered Object-Glass.—Herschel's Solar Eyepiece.—The Polarizing Eyepiece.

The heat and light of the sun are so intense that peculiar instruments and methods are necessary for the observation of his surface. The appliances used in the study of the moon, planets, and stars will not answer at all for solar work.

A very excellent method of proceeding where the object is to secure a general view of the sun, without regard to delicate detail, and to determine easily and rapidly the positions of spots and other objects on the sun's disk, is to project his image upon a sheet of cardboard by means of a telescope.

For this purpose things are arranged as indicated in the figure. The sheet of paper upon which the image is to be thrown is supported in front of the eyepiece by a light framework attached to the telescope. The distance of the screen from the eyepiece depends upon the size of image desired and the power of the eyepiece; a diameter of from six inches to a foot being generally most convenient. Another screen is usually fitted on the object-glass end of the telescope to balance the first, and shade it from all light except that which has passed through the instrument. If the apparatus is to be used to determine the position of spots on the sun, the surface which receives the image must be carefully adjusted so as to be perpendicular to the optical axis of the telescope.

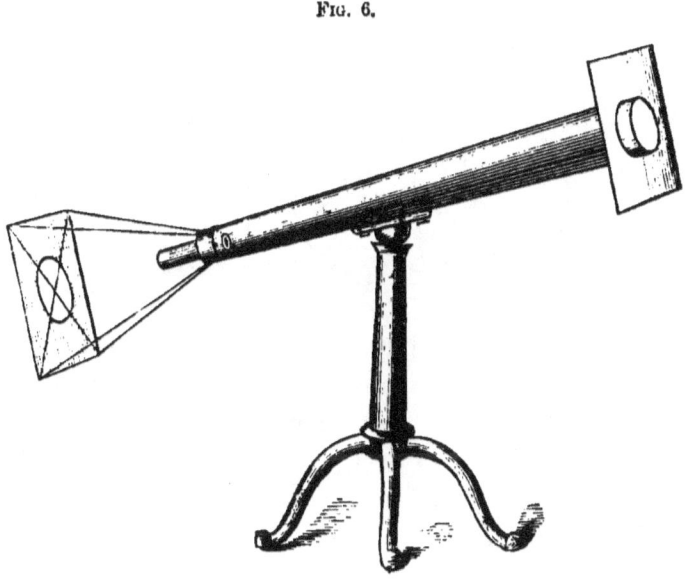

Fig. 6.

To determine the position of objects on the sun's disk, Carrington used two lines, ruled at right angles to each other upon the screen, and set at an angle of about 45° with the north and south line or hour-circle. The observations needed to determine the place of a spot on the sun's disk then consist merely in noting with a watch as accurately as possible the four moments at which the edge of the sun's image crosses the two lines (the telescope being, of course, firmly fixed during the whole time), and the two moments when the spot passes them. From these six observations, with the help of the data given in the almanac, the distance and direction of the spot from the sun's center may readily be calculated by formulæ which would hardly be suited to these pages, but which may be found in the monthly notices of the Royal Astronomical Society, vol. xiv, page 153. Fig. 7 illustrates this arrangement.

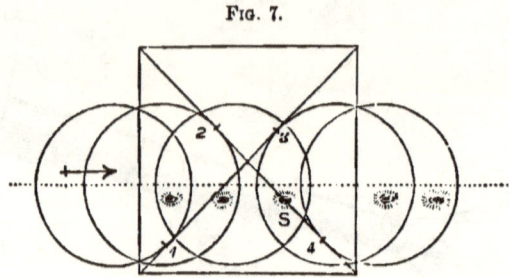

Fig. 7.

A simple method, but not so uniformly accurate, is to rule upon the screen a circle whose diameter is best about half that of the field of view, and then note the instants when the edge of the solar image is tangent to this circle, and the two moments when the spot crosses it. With an equatorial instrument (that is, one so mounted that its principal axis of motion points to the celestial pole) Carrington's method is preferable, since the lines once adjusted to the proper position retain it in all positions of the telescope. With an ordinary telescope, not so mounted, the circle is more convenient.

With a small telescope thus fitted up, one is in a position to make observations of real value as to the number, position, and motions of the solar spots. Occasionally, also, when the air happens to be in good condition, a considerable amount of detail can be made out by this method in the spots and upon the solar surface generally. The darkening of the edge of the sun, caused by the absorption of the solar atmosphere, is very noticeable, and the faculæ are conspicuous. One great advantage of the method is, of course, that several persons can thus observe together. A teacher, for instance, can in this way exhibit to a class of a dozen all the principal features of the sun's surface, and be sure that they all see the things he desires them to notice.

Should any amateur happen to find upon the sun's disk a small, round spot, which he has reason to think is an intra-Mercurial planet, a few observations of the sort indicated above, repeated at intervals of some minutes, would settle the question immediately, and give a reasonably accurate determination of the rate and direction of movement.

If the instrument has an equatorial mounting and clockwork, so that the image remains apparently stationary upon the screen, a very satisfactory tracing can be made upon paper ruled in squares, showing pretty accurately the position and magnitude of all visible spots, in a form suitable to file away for reference. The observations of Carrington's great work upon the solar spots were for the most part made in this manner.

Of late years photography has been extensively utilized for observations of this sort. The apparatus consists of a telescope fitted with a camera-box in place of an eyepiece, and with an arrangement for producing an instantaneous exposure of the sensitive plate to the solar rays.

Since, in the ordinary achromatic telescope, the rays which are most effective in photographic action do not come to a focus at the same point as those which most strongly affect the eye, such an instrument, however perfect visually, will not give sharp photographic impressions. It is necessary, for the best photographic results, to use object-glasses whose corrections are calculated expressly for the purpose. Mr. Rutherfurd, of New York, seems to have been the first to appreciate this, and to construct an instrument specially designed for astronomical photography. To this end, disdaining all compromise, he did not hesitate to sacrifice deliberately the visual excellence of an exquisite object-glass of thirteen inches diameter, by altering its curves so as to produce the most perfect actinic correction; and he has been rewarded by a success hitherto unequaled as regards the perfection of the pictures obtained. Some of his photographs of the sun and moon rival in sharpness and detail the drawings of accomplished observers.

Another and simpler method of obtaining the desired corrections, originally tried by Mr. Rutherfurd and rejected as not absolutely the best possible, has recently been revived by Cornu, of Paris. It consists, not in regrinding the two lenses which compose the object-glass, but merely in separating them slightly—half an inch or so for an instrument of ten-feet focus. The approximate correction, thus produced, gives excellent results, and the instrument is not spoiled for other work, since it requires only a few minutes to restore the glasses to their visual adjustment.

In a reflecting telescope there is, of course, no difficulty of this sort, since rays of different wave-length and color are not dispersed by reflection as by refraction. Other and still more serious difficulties, however, exist, depending upon the extreme sensitiveness of the reflector to the distorting influence of variations of temperature; so that, hitherto, reflectors have not equaled refractors in the excellence of their photographic work. They have been employed with very good success, how-

METHODS FOR STUDYING THE SURFACE OF THE SUN. 55

ever, on several occasions for the photography of solar eclipses.

With telescopes of considerable size the picture is generally formed directly at the focus of the object-glass without further enlargement. This is the case with the pictures made by Mr. Rutherfurd, in which the diameter of the sun's image is about 1¾ inch..

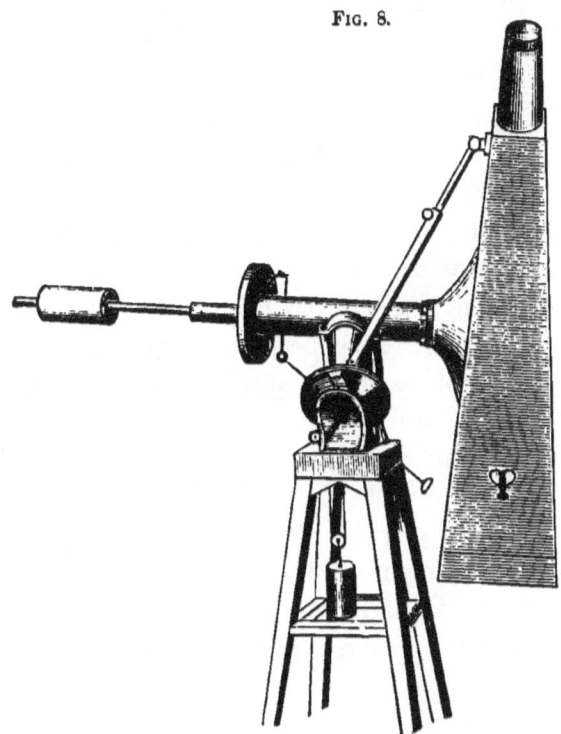

Fig. 8.

KEW PHOTOHELIOGRAPH.

Copies of the negatives are afterward made if desired on a larger scale. In smaller instruments, such as the well-known photoheliograph of the Kew Observatory, an enlarging eyepiece is used, so constructed as to distort as little as possible the image formed by the object-glass while magnifying it to a diameter of three or four inches. In this instrument, of which we give a figure, the diameter of the object-glass is only 3½ inches, and its focal length 50 inches; the tube, instead of being conical as usual and larger at the object-end, is made pyramidal and larger at the bottom, in order to accommodate the plate-holder more conveniently. The whole is mounted equatorially, and driven by clockwork. It was constructed in 1857, under the directions and after the designs of Mr. De La Rue, at the request of the council of the Royal Society, and has proved itself a most efficient and excellent instrument. A number of other very similar instruments have since been made with slight improvements. Those employed by the English and Russian parties in their photographic operations at the last transit of Venus, were of this type. So also were those of the German parties, except that they had considerably larger telescopes, with apertures of from six to eight inches, and therefore needed less powerful enlarging lenses. As has been mentioned in another connection, the French and Americans used a different arrangement, employing object-glasses with a focal length sufficiently great (from twenty to forty feet) to render enlargement of the image unnecessary, placing the tube horizontal, and reflecting the light through the lens with a plane mirror moved by clockwork.

The sunlight is so powerful that the exposure of the plate has to be made practically instantaneous. The apparatus by which this is effected varies greatly in detail in instruments of different types, but in all cases consists essentially of a slide, carrying in it a slit of adjustable width and capable of being shot across in front of the sensitive plate by a strong spring. At the moment of exposure a trigger or telegraphic key is

METHODS FOR STUDYING THE SURFACE OF THE SUN. 57

touched by the operator, and the slide, previously drawn back and locked by suitable mechanism, is released, and in its flight allows the rays to gleam through the aperture for a time, which in different instruments varies from $\frac{1}{100}$ to $\frac{1}{5000}$ of a second, according to the size of the instrument, the sensitiveness of the collodion, and the clearness of the atmosphere.

We give a figure of Vogel's exposure-slide, which is perhaps as good as any. M is an electro-magnet, which,

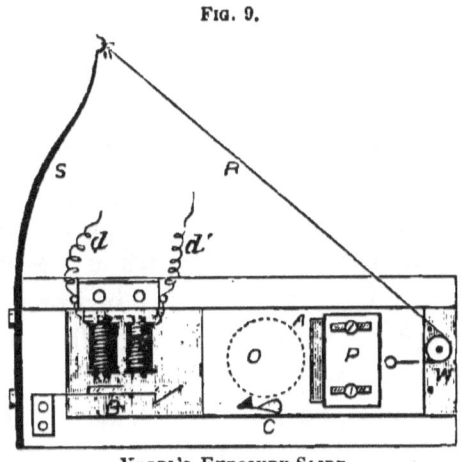

FIG. 9.

VOGEL'S EXPOSURE-SLIDE.

on the touch of a telegraph-key in the observer's hand, attracts the armature B, thus releasing the catch C, and allowing the spring S, by the intervention of the cord and pulley, to draw the slide containing the slit A swiftly across the orifice through which the rays enter the camera.

The character of the picture produced depends very greatly upon the proper timing of the exposure. If the intention be to secure an image of the sun with hard, firm edges from which measurements can be made to determine the position of objects on the solar disk—

as was the case at the transit of Venus—then a relatively long exposure is needed; but it is to be remembered that the diameter of the sun's image increases very perceptibly with lengthening exposure, so that this diameter can never be safely used to furnish the scale of measurement. If, on the other hand, what is desired is a picture full of detail, showing the faculæ and the structure of the spots, the exposure must be greatly shortened by narrowing the slit or giving the slide a greater velocity; and it must be added, unfortunately, that the exposure which brings out perfectly the central portions of the disk is altogether too short for the portions near the limb, where the actinic power is very greatly diminished.

This circumstance detracts considerably from the value of the photographic method. The skillful draughtsman can show in the same picture details differing to any extent in intensity, while the photograph is, so to speak, limited to the reproduction of only one certain class of details at a time. Still we can always be sure that, whatever a photograph does show, is an autographic representation of fact, and not a figment of the imagination. This is not the case with drawings; for it is remarkable how widely two conscientious artists will differ in their representations of the same object, seen by both with the same telescope, and under the same circumstances. As an accurate record of the number, position, and magnitude of the solar spots at any given time, the photograph is, of course, unexceptionable. Such a record has been obtained by the Kew photoheliograph for fourteen years—from 1858 to 1872. In 1872 the work at Kew was given up, and the instrument transferred to Greenwich, where since the beginning of 1873 the series has been continued, at least two pictures being taken every day when the weather will permit, and more than two if anything of especial interest is going on upon the solar surface. A similar series was kept for many years at the observatory of Wilna in Russia, until it was destroyed by fire in 1877. The new physical observatories of France and Germany propose the same thing. Since it is quite possible that clouds may cover all these stations at once, it seems very desirable that instruments of the same sort should be established on the Western Continent, and in the southern hemisphere, so as to secure for astronomy a practically continuous record.

Very recently, Janssen, at the new French physical observatory at Meudon, has carried solar photography to a point far beyond any previous attainment. He has accomplished it mainly by utilizing the fact that there exists in the spectrum, near the Fraunhofer line G, a narrow band of rays which possess a photographic activity upon the salts of silver much more intense than that of any other portion of the spectrum. It is so intense, indeed, that if the exposure be very short and properly regulated, the effect is practically the same as if the sunlight were monochromatic, consisting of these rays alone: any defect in the color-correction of the object-glass is rendered almost harmless. This makes it possible to use an ordinary achromatic object-glass, roughly corrected for photographic work by merely separating the lenses a trifle, according to Cornu's plan.

With a five-inch telescope and a suitable enlarging lens, Janssen produces pictures even half a metre in diameter, and of extreme perfection in their delineation of the details of the solar surface. The exposure, ranging from $\frac{1}{200}$ to $\frac{1}{1000}$ of a second, according to the clearness of the air and the altitude of the sun, is effected by a slide closely resembling Vogel's. The impression obtained is very feeble, and requires prolonged and careful development; but, when at last fairly brought out, is every way admirable. Some very interesting results, which we shall deal with later, have already been deduced from his plates.

Photography, however, is not adequate as yet to the study of the most delicate details of the solar surface. For this purpose nothing can take the place of ocular observation by experienced and skillful observers, armed with powerful telescopes and suitable appliances, and on the watch for the few favorable moments when the atmospheric conditions will permit successful work.

The instrument must be provided with some form of solar eyepiece expressly adapted to the purpose. The old-fashioned way was to use an ordinary eyepiece, fitted with a dark glass next the eye. If the whole aperture of a telescope of any size is used, the heat at the focus is so great as to endanger the lenses, and accordingly it was customary to "cap down" the object-glass—i. e., to put on a cover with a small hole in the center, so as to reduce the aperture to two or three inches. In this way, of course, the heat and light are easily diminished to almost any extent, but the definition is greatly injured. According to well-known optical principles, the image of a luminous point is not a point, even in an absolutely perfect telescope, but, in consequence of the so-called "diffraction" due to the interference of light, becomes a small disk, surrounded by a series of concentric luminous rings; the smaller the aperture of the telescope, the larger the disk with a given magnifying power. Similarly, the image of a luminous line is not a line, but a stripe of determinate width with fringes on each side. It is easy to see, therefore, that a telescope of small aperture can not possibly be made to show as delicate details as one of larger diameter, and, to get the best results in examining the surface of the sun, we must find some way of diminishing the light and heat without cutting down the diameter of the object-glass (or mirror, if we are using a reflecting telescope).

A reflecting telescope whose mirror is of *unsilvered glass* effects this very beautifully. The unsilvered surface reflects only about $\frac{1}{80}$ of the incident light and heat, and although the resulting image is still too bright for the unprotected eye, the heat is not troublesome, and only a very thin shade-glass is needed. Another excellent method is to silver by Liebig's or some analogous process the front surface of the object-glass of a refractor. The silver film can be deposited of such a thickness as to allow any desired percentage of the light to pass, while the rest is reflected and not allowed to enter the instrument at all. The image formed in this way is slightly tinged with blue, but is beautifully sharp and steady, there being a great advantage in preventing the heating of the air in the telescope-tube, which occurs with every other form of instrument. The telescopes employed by the French parties in the observation of the late transit of Venus were prepared in this way. With its great advantages, however, the method has on the whole quite as great disadvantages, as was evident at Saigon, where clouds were so thick that nothing could be seen through the silver film, and the observer had to rub it off with a cloth before he could do anything. Then, of course, a telescope prepared in this way can not be used for any other purpose. The common practice, therefore, is, not to adapt the instrument for solar observation by doing anything to its object-glass or mirror, but to accomplish the desired result by some modification or accessory of the eyepiece.

One of the best known and most generally useful eyepieces is that devised by Sir John Herschel, and bearing his name. It is represented in Fig. 10, which gives a section of it. The light entering at O encoun-

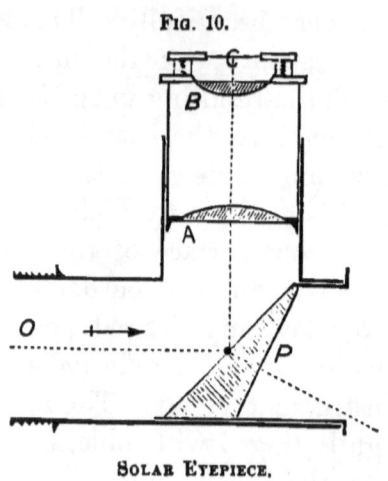

SOLAR EYEPIECE.

ters a prism of glass, whose first surface is placed at an angle of 45°. The greater part of the light, something over $\frac{19}{20}$, passes through the prism, emerging perpendicular to its second surface, and goes out through the open end of the tube; the reflected light, about $\frac{1}{20}$ of the whole, is thrown upward through the eyepiece proper, A B, which is precisely the same as ordinarily used. In this way most of the light and heat are got rid of; too much, however, still passes the lenses for the eye to bear, and it is necessary to use a shade-glass; but this may be very light. The brightness of the sun varies so much at different altitudes and under different conditions of the atmosphere, that it is desirable to have the thickness of the shade-glass adjustable. This is easily managed by using a long, thin wedge of dark glass, compensated by a corresponding wedge of ordinary glass, and set in a proper frame, as represented in Fig. 11. The shade-glass should not be

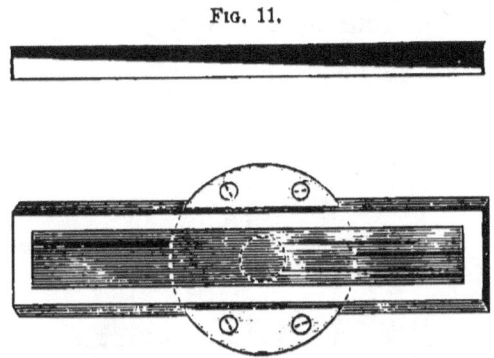

FIG. 11.

colored, but of neutral tint, so that objects on the sun's surface may be seen of their proper hue. The glass known as "London smoke" very nearly fulfills this condition, and with a shade of this material the apparatus is exceedingly satisfactory, and quite sufficient for ordinary work.

Still finer results, however, may be obtained with more complicated and expensive "helioscopes," as they are called, which by means of polarization reduce the light to such a degree that no shade is needed, and, moreover, enable us to graduate the light as we please by merely turning a milled head. There are several forms of the apparatus: we give a figure of one constructed by Merz, slightly modified,* which is perhaps as convenient and effective as any. The light entering at A first en-

---

* The modification consists in substituting the prisms P^1 and P^2 for simple reflectors of black glass, which are very apt to be broken by the heat of the sun's image.

counters the surface of a prism, P^1, set at the polarizing angle; about $\frac{15}{16}$ of the light passes through the prism, emerging perpendicular to its rear surface, and, being rejected, about $\frac{1}{16}$ is reflected and polarized by the reflection. The reflected ray next strikes the surface of a second prism, P^2, and here a considerable portion of the remaining light is thrown away. That which

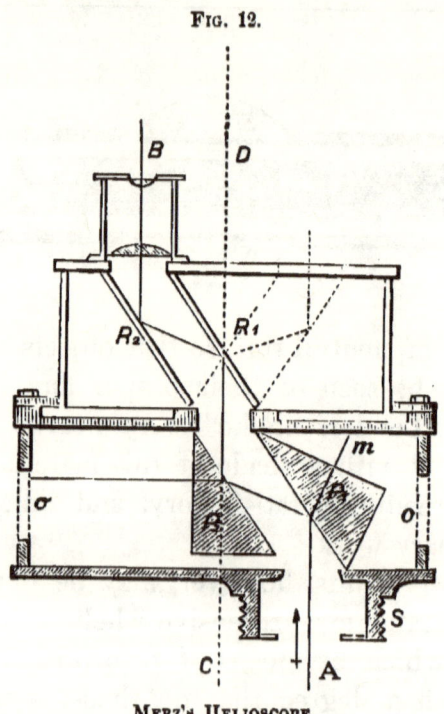

Fig. 12.

MERZ'S HELIOSCOPE.

is left is reflected into the upper portion of the eyepiece parallel to its original direction, through an opening in the top of the circular case in which the two prisms are mounted. The upper case is attached to the lower in such a manner that it can be turned around the line C D as an axis. It contains two plane mirrors of black glass, placed as shown in the figure. With things in the position indicated, a beam of considerable strength would reach the eye at B—so strong, in fact, as to be painful; and the same would be the case if the upper piece were turned 180°, bringing the mirrors into the position shown by the dotted lines, with the issuing ray in the prolongation of the incident. But, by turning the upper piece one quarter of a revolution, the issuing ray can be entirely extinguished, and, by turning it less or more than 90°, the intensity of the light can be controlled at pleasure. As no shade-glass is used, everything is seen of its proper tint. Another advantage is, that there is no such disturbance of the orientation of the solar image as happens with every form of diagonal eyepiece. North, south, east, and west fall in their usual and natural places—a matter of some importance as regards the convenience of observation.

Still other forms of helioscopic eyepiece depending upon polarization have been devised by Secchi, Langley, Christie, Pickering, and others, each with its own peculiar advantages; our limits, however, forbid more extended treatment of the subject. We add merely that in some cases, as in the study of the internal structure of sun-spots, it is found very advantageous to adopt the device of Dawes, and limit the field of view by a minute diaphragm made by piercing a card or plate of ivory with a hot needle; thus excluding the light from any portion of the sun's surface except that under immediate observation.

CHAPTER III.

THE SPECTROSCOPE AND THE SOLAR SPECTRUM.

The Spectrum and Fraunhofer's Lines.—The Prismatic Spectroscope; Description of Various Forms and Explanation of its Operation.—The Diffraction Spectroscope.—Analyzing and Integrating Spectroscopes.—The Telespectroscope and its Adjustment.—Explanation of Lines in the Spectrum.—Kirchhoff's Researches and Laws.—The Sun's Absorbing Atmosphere and Reversing Layer.—Elements present in the Sun.—Lockyer's Researches and Hypothesis.—Basic Lines.—Dr. H. Draper's Investigations as to the Presence of Oxygen in the Sun.—Schuster's Observations.—Effect of Motion upon Wave-Length of Rays and Spectroscopic Determinations of Motion in Line of Sight.

EVER since the time of Newton it has been known that a beam of white light is decomposable into its constituent colors by passing it through a prism, and, under certain circumstances, the result is a rainbow-tinted band or ribbon, which has been called the solar spectrum. In this spectrum Wollaston, in 1802, discovered certain dark shadings, and in 1814 Fraunhofer again and independently discovered the same thing; and he so improved his apparatus and method of observation as to get not merely indefinite shadings, but clear, sharp lines, of which he made a map, assigning designations to many of the principal ones. Indeed, these markings of the solar spectrum bear his name to this day.

He, however, could not account for them, further than to show that they did not originate in his instrument nor in the earth's atmosphere; and it was not until the publication of the researches of Kirchhoff and Bunsen, in 1859 and 1860, that the scientific world came to appreciate their meaning and importance.

We speak of the work of Kirchhoff and Bunsen as epoch-making, and such was certainly the case. At the same time the secret of the solar spectrum had been, in part at least, divined before by Stokes, Thomson, and Ångstrom; the latter especially, whose memoir, published in 1853, would certainly have obtained a high celebrity if it had appeared in French, English, or German, instead of Swedish. Swan and Zantedeschi had also given to the spectroscope nearly its present form; and a number of other investigators, among whom Sir John Herschel, Wheatstone, Foucault, and J. W. Draper deserve special mention, had each contributed something important to the foundations of the new science, for such it has proved to be. The study of spectra has opened a new world of research, and added some such reach to our physics and chemistry as the telescope brought to vision.

Of course, any extended discussion of the instruments, principles, and methods of spectroscopy would be inconsistent with our limits: we can only treat the subject very briefly.

First, then, as to the instrument. It consists usually of three parts: the collimator so called; the light-analyzing apparatus, which is sometimes a prism or train of prisms, and sometimes a diffraction grating; and the view-telescope. The figure shows the construction of a single-prism spectroscope, and the course of the rays of light through it. The collimator is simply a telescope without an eyepiece, and having in the place of the eyepiece a narrow slit. This slit is placed exactly at the focus of the object-glass of the collimator, so that rays from each point of the slit become parallel beams after passing the lens, and a person looking through the object-glass, at the slit, sees it precisely as if it were an object in the sky. Optically, the slit of the collimator

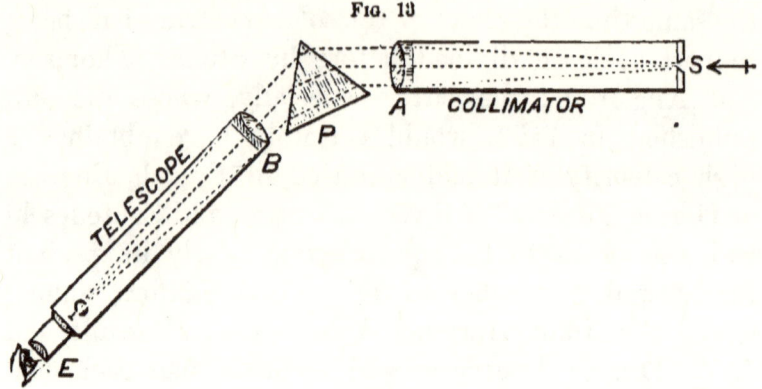

ARRANGEMENT OF PRISMATIC SPECTROSCOPE.

is thus removed to an infinite distance; while, mechanically, it is still at the fingers' ends, within reach of manipulation and adjustment. The collimator, however, is not essential. Fraunhofer's work was all done with light admitted through a slit in the window-blind at a distance of twenty or thirty feet—a much less convenient arrangement, as is at once evident.

The view-telescope, which, however, is no more essential than the collimator, is usually a small telescope with an object-glass of the same size as that of the collimator, and magnifying from five to twenty times. Generally, the collimator and telescope of astronomical spectroscopes are from three quarters of an inch to an inch and a half in diameter, and from six to eighteen inches long.

The light, after passing the slit and object-glass of the collimator, next strikes the prism or grating, and these two things—the slit and the prism or grating—

are really all that is essential. In the case of a prism the rays are bent out of their course, as shown in the figure, and enter the view-telescope, which is placed at the proper angle to receive them. Suppose, now, for a moment, that the light admitted at the slit is strictly homogeneous—say red. The eye at the view-telescope would then see a red image of the slit, corresponding precisely in form and proportions to the slit itself, widening if the slit is widened by its adjusting screw, or narrowing down to a mere line if the jaws of the slit are screwed up close. If instead of a slit the opening had some other form, as an arc of a circle, a triangle, or a square, the image seen would imitate it, always having the same color as the light admitted. Suppose, again, that the light is not homogeneous, but consists of two kinds mixed together—say red and yellow. Viewing the slit directly, without the spectroscope, one would only see a single orange-colored image; but with the spectroscope one would see *two* widely separated images, one of them red, the other yellow. This is because the prism refracts the two kinds of light differently, so that after the rays have passed the prism they strike the object-glass of the view-telescope in different directions, and then make images in different places. If the light is composed not of two kinds only, but many, the images will be numerous, ranged side by side like the pickets of a fence; and if, as in the case of a candle-flame, the light emitted contains an indefinite number of tints, then the slit-images, placed side by side, will coalesce into a continuous band of color. If, in the candle-light, certain kinds of light are specially abundant, then the corresponding slit-images will be more brilliant than their neighbors; and if, as is usually the case, the slit be narrowed to a line, these slit-images will become *bright lines* in the spectrum—*lines* only because the slit is itself a line, which, of course, is the best form to give the light-admitting aperture, in order that the different images may overlap and interfere as little as possible.

If any kinds of light be wanting, then the corresponding images of the slit will be missing, and the spectrum will be marked by dark bands or lines.

FIG. 14.

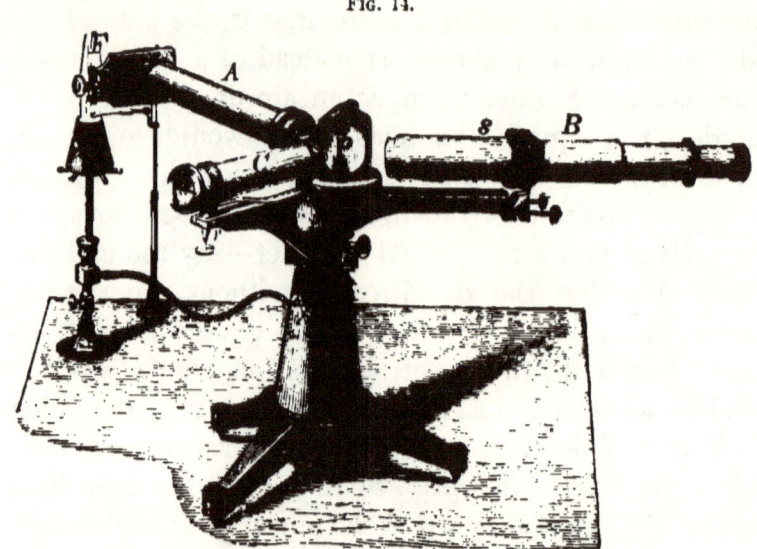

BUNSEN'S SPECTROSCOPE.

The cut (Fig. 14) shows the actual appearance of what is known as the chemical spectroscope, ordinarily used in laboratories. Besides the collimator A, and the telescope B, it has a third tube C, which carries a fine scale photographed on glass at the end farthest from the prism. There is a lens in the tube at the end next the prism, so that the observer at the telescope sees this scale running across the field of view at the edge of the spectrum, and thus has the means of noting accurately the position of any lines he may find. This arrangement is due to Bunsen.

It is often desirable to obtain a greater separation of the different colors—dispersion, to use the technical term—than a single prism would produce. In this case, the rays after passing through the first prism may be transmitted through a second and a third, and so on, until they reach the view-telescope. With prisms as commonly made, it is difficult to use more than six in this manner, but it is possible by reflection properly managed to return the rays through a second prism-train connected with the first, so as to get the virtual effect of from ten to twelve prisms. The instrument figured on page 78, and used for observation of the solar prominences, is of this kind.

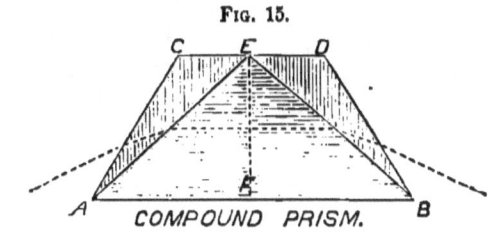

FIG. 15.

COMPOUND PRISM.

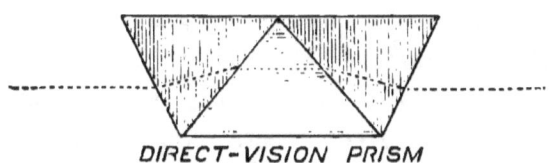

DIRECT-VISION PRISM

Another way is to use a compound prism, so called, each composed of a very obtuse-angled prism, A B E, of some highly dispersive material, usually heavy flint glass, flanked by two prisms of lighter glass with their refracting angles reversed. Prisms of this kind can be made of much higher dispersive power than simple prisms, and of course a smaller number will answer the same purpose. By properly proportioning the angles C A E and E B D, it is possible to make the yellow rays of the spectrum pass through without change of direction, while still retaining a considerable dispersion. An instrument with prisms of this kind is called a "direct-vision" spectroscope, and in some cases is much more convenient than the other forms.

Thollon has recently constructed compound prisms having the dense glass prism replaced by a chamber filled with carbon disulphide, which possesses an enormous dispersive power; with a train of these prisms he has obtained views of the spectrum only equaled by the performance of the best diffraction gratings. A dispersion equal to that of thirty or forty prisms of an ordinary spectroscope is easily reached. The behavior of these disulphide prisms is, however, far from satisfactory for ordinary work, since they are extremely sensitive to small changes of temperature, which cause irregular refractions in the liquid, and destroy the definition.

We have used the expression, the dispersive power of thirty or forty prisms; but that is very indefinite, because the dispersive power of a spectroscope depends upon its linear dimensions as well as the kind and number of prisms and is proportional to the dimensions. That is to say, if a given spectroscope has the size of its prisms, and the diameter and focal lengths of its collimator and telescope doubled, retaining, however, the former slit and eyepiece, its dispersive power will be doubled by the change. Thus a large single-prism instrument may equal in working power a small one of many prisms. Lord Rayleigh has recently shown that the *resolving* power of a spectroscope, constructed with prisms of a given substance, depends upon the length of the route pursued by the rays of light in traversing them.

As has been said, a diffraction grating may replace the prism in a spectroscope. This diffraction grating is merely a system of close, equidistant, parallel lines ruled upon a plate of glass or polished metal. The best hitherto made are those produced by Mr. Chapman upon a machine constructed for the purpose by Mr. L. M. Rutherfurd, of New York.

One of these gratings in the writer's possession is ruled upon speculum metal, the lines being each an inch and three quarters long. The ruling covers a space of more than two inches, the interval from each line to the next being $\frac{1}{17280}$ of an inch, and the whole number nearly forty thousand. The closer the lines, the greater the dispersion produced; the larger the ruled surface, the more light is at the observer's disposal, provided the collimator and view-telescope are large enough to utilize the whole ruling. The greater the total number of lines, the higher the resolving power of the grating, or power of separating close lines in the spectrum.

An explanation of the manner in which the grating operates to produce its spectra would take us too far; for this we must refer the reader to any good treatise on optics. We say *spectra*, because, while a prism gives but one spectrum, a grating gives many, and of different degrees of dispersion, which is often a matter of great convenience. Of course it will be easily understood that no one of the spectra is as brilliant as if it were the only one.

The grating mentioned above, in combination with

a collimator and telescope of about four feet focal length, exceeds, or at least equals, in spectroscopic power, anything ever yet constructed, and in convenience is incomparably superior to any instrument with a train of prisms.

Fig. 16 shows the arrangement of the different parts of such an instrument. The collimator and view-tele-

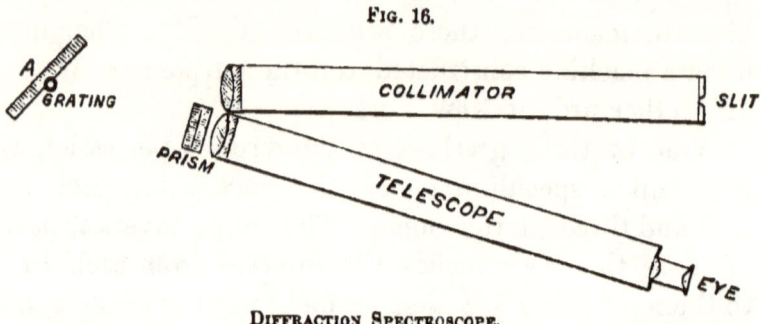

FIG. 16.

DIFFRACTION SPECTROSCOPE.

scope are placed with their object-glasses close together, and their tubes making as small an angle as possible, consistently with keeping the grating at a manageable distance. Of course, collimator and telescope must both be pointed at the center of the grating. The grating is mounted on a frame with an axis at Λ, so that it can rotate in the plane of the dispersion, the ruled lines being parallel to this axis. The frame which carries the grating must be so constructed as to support it steadily and firmly, without the slightest strain, for it is essential to its good performance that the surface be strictly plane. Although the grating just mentioned is ruled upon a plate of speculum metal only three inches square, and nearly three eighths of an inch thick, an abnormal pressure of a single ounce at one of the corners will sensibly affect its performance, and four ounces bends the plate sufficiently to ruin the definition.

As the different orders of spectra overlap each other (the red end of the second order spectrum overlapping the blue of the third, etc.), it is sometimes necessary to separate them, and this can be done in a manner first suggested by Sir John Herschel, by interposing between the grating and view-telescope a single prism with its plane of dispersion perpendicular to that of the grating, the telescope being then inclined at the proper angle to receive the rays. A direct-vision prism in the eyepiece answers the same purpose, though less satisfactorily. In many cases a suitably colored shade-glass is sufficient.

FIG. 17.

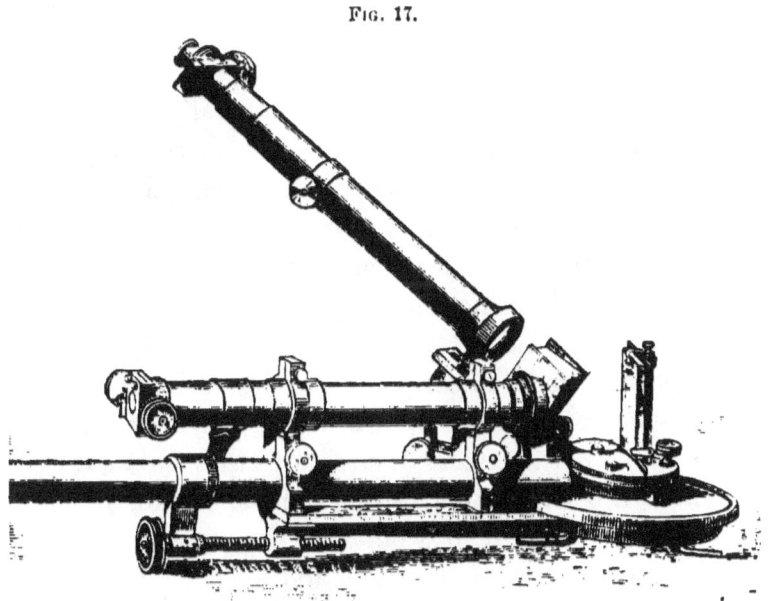

PRINCETON SPECTROSCOPE.

Fig. 17 is from a photograph of the instrument actually in use at Princeton for observations upon solar prominences. It is designed to be attached to the equatorial, and is therefore on a smaller scale than the one mentioned above, its collimator and view-telescope being each only about thirteen inches long, with a diameter of an inch and a quarter. It uses the same grating, however.

The prismatic and diffraction (or interference) spectra differ from each other to a certain extent, not of course in the order of colors or of lines, but in their relative distances. In the prismatic spectrum the red and yellow portion is much compressed, while the violet is greatly extended; with the diffraction spectrum the reverse is the case; the lines in the violet are crowded together, and those in the red are widely separated.

In the diffraction spectrum the lines are perfectly straight; in the prismatic, generally more or less curved; we say generally, because there are forms of high-dispersion spectroscope in which this curvature is corrected. This curvature is caused by the fact that the rays from the top and bottom of the slit do not meet the refracting surface at the same angle as those from the middle of the slit; they are, therefore, differently refracted; in consequence, the slit-images of which the spectrum is built up are not straight but distorted.

It is hardly necessary to add that the dark lines which run lengthwise through the spectrum are merely due to particles of dust between the jaws of the slit. It is almost impossible to make and keep the edges of the slit so clean and smooth that lines of this sort will not appear when the opening is made very narrow.

The spectroscope may be used in two entirely different ways: it may simply have its collimator pointed toward the source of light; or a lens may be interposed between the slit and the luminous object, so as to form an image of the latter on the slit.

In the first case, the instrument is said to be an *in-*

tegrating spectroscope, because each point in the slit receives light from the whole of the luminous object, so that the spectrum is alike through its whole width, and represents the average light of the object—it *lumps* the whole, so to speak. In the second case, different parts of the slit are illuminated by light from different parts of the object; the top of the slit gets the light from one point, the middle of the slit from another, and the bottom from a third. If, then, the lights emitted by the three points differ, their spectra will differ also, and the observer will find that different portions of the width of his spectrum will differ correspondingly—the upper portion will be unlike the middle, and the middle will differ from the bottom. An instrument arranged thus is called an *analyzing* spectroscope, because it enables us to determine separately the spectra of various portions of an object, and thus to analyze its constitution; as, for instance, a sun-spot and its surroundings. For most purposes, especially astronomical, it is much the most satisfactory. Approximately the same end may be reached, in some cases, by placing the slit very near the luminous object, as in flame analysis, but it is usually much more convenient and better to use the lens. In astronomical work the object-glass of a large equatorial telescope is generally employed to form the image of the celestial object, and the spectroscope is attached at the eye-end of the telescope, the eyepiece being removed. The combined instrument is then often called a tele-spectroscope. The figure on the next page represents the apparatus used at the Dartmouth College Observatory.

It is usually very important that the slit of the instrument be precisely in the focal plane of the object-glass of the telescope for the rays specially under examination. On account of the so-called "secondary spectrum" of the achromatic lens, this focal plane is quite different for the different colors, and the spectroscope requires to be slid in or out, so as to vary the distance of its slit from the great object-glass of the

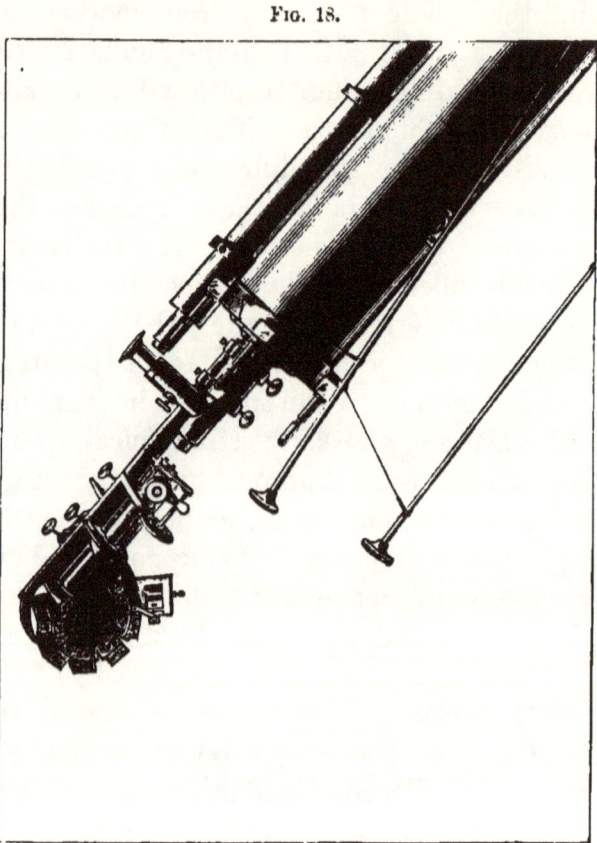

FIG. 18.

TELE-SPECTROSCOPE.

telescope according to circumstances. The same end may be obtained (less satisfactorily, however) by a second lens between the object-glass and the slit, and pretty near the latter. By moving this lens, the focus can be made to fall exactly on the slit. Neglect of this adjustment will make many of the most interesting and important spectroscopic observations quite impossible.

If the collimator of a spectroscope of any form be directed toward an ordinary lamp, or upon the incandescent lime of a calcium-light, the observer will get simply a continuous spectrum; a band of color shading gradually from the red to the violet, without markings or lines of any kind. If the instrument be turned toward the sun he will obtain something much more interesting—a band of color, as before, but marked by hundreds and thousands of dark lines, some fine and black, like hairs drawn across the spectrum, while others are hazy and indistinct.

Most of them retain their appearance and position perfectly from day to day; some of them, however, are more intense at one time than another, and when the sun is near the horizon certain lines in the red and yellow become extremely conspicuous, in such a way as to make it clear that they, at least, have something to do with our terrestrial atmosphere. Fig. 19 is a reproduction of a

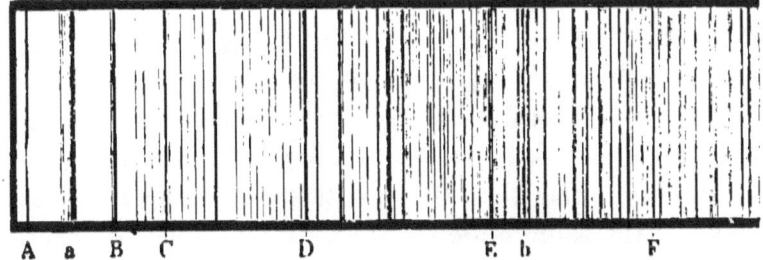

FIG. 19.

portion of Fraunhofer's map of the solar spectrum, showing what one might fairly expect to see (except as to color) with an excellent single-prism spectroscope. Fig. 20 is a drawing of a very small portion of the spectrum in the green, as shown by the powerful diffraction spectroscope mentioned a few pages back. The scale is that of Ångstrom's map. The large, heavy lines are

FIG. 20.

b GROUP IN SOLAR SPECTRUM.

known as the *b* group, and are due, as we shall soon see, part of them to the presence of iron and nickel, and part to magnesium, as *gases* in the solar atmosphere.

If, instead of using the sun or an ordinary flame for the source of light, we examine with the spectroscope an electric spark, or the arc between carbon points, or the light produced by passing the discharge of an induction coil through a rarefied gas, we shall get a spectrum of quite a different sort—a spectrum consisting of bright lines upon a dark or faintly luminous background; and it will be found that the spectrum developed will always be the same under similar circumstances, depending mainly upon the material of the *electrodes* (the points between which the discharge passes), and the nature of the intervening gas, but also, to a certain extent, upon its density and the intensity of the electric discharge. So, also, if certain easily vaporized salts are introduced into the blue flame of a Bunsen gas-burner, or of a spirit-lamp even, the flame becomes colored, and its spectrum is a spectrum of bright lines, which are perfectly characteristic of the metal whose salt is used. An ordinary candle-flame, indeed, almost always shows one such bright line in the yellow, as had been noticed many years before Swan, in 1857, showed it to be due to the presence of sodium, which in the form of common salt is universally distributed.

Fraunhofer, as early as 1814, had discovered that this line (or lines rather, for it is really composed of two, easily separated by a spectroscope of no great power) exactly coincides with the double line which he named D, in the solar spectrum; and he had found the same line in the spectra of certain stars also; but he did not know that the line was due to sodium, or in all probability he would have anticipated by nearly half a century the discovery which lies at the foundation of modern spectrum analysis. As has been said before, the principles involved seem to have been more or less distinctly apprehended by several persons—Foucault and Ångstrom especially—years before the publication of Kirchhoff in 1859; but it was his work which first bore fruit.

It is not necessary to repeat here again the oft-told story how he found that, when sunlight is made to pass through a flame containing sodium-vapor, the D-lines in the spectrum of this sunlight come out with increased intensity; though, when a screen is interposed between the sun and the flame, the lines are bright, as usual in such a flame. He found, too, that when the incandescent lime-cylinder of the calcium-light is placed behind the sodium-flame, a precisely similar phenomenon occurs, and the bright lines of the flame-spectrum are reversed to dark ones.* He found the same thing to hold good also for a flame colored by lithium.

The sum of his results may be stated as follows:

1. Solids and liquids, when incandescent, give continuous spectra; and, as we now know, the same thing is true of gases also at great pressures.

2. Bodies in the gaseous state (and not compressed) give discontinuous spectra consisting of bright lines and bands; and these bright-line spectra are different for different substances and characteristic, so that a given substance is identifiable by its spectrum.

3. When light from a solid or liquid incandescent body passes through a gas, the gas absorbs precisely those rays of which its own spectrum consists; so that the result is a spectrum marked by *black* lines occupying exactly the same positions which would be held by the bright lines in the spectrum of the gas alone.

If, then, sodium is present in the solar atmosphere between us and the photosphere, we ought to find in

---

* The blackness of the lines formed in this way is such that it is sometimes difficult to believe, what is really the fact, that they are actually *brighter* than they were before the lime-cylinder was placed behind the flame, and that their darkness is only apparent, and due to their contrast with the more brilliant background of the continuous spectrum of the incandescent lime. It is very easy, however, to demonstrate the truth by a simple experiment. Insert in the eyepiece of the view-telescope of a spectroscope of some power, an opaque diaphragm pierced with two slits at right angles to each other, thus, $(a) \frac{(b)_|}{x}$. Put before the collimator slit a sodium-flame, and, by a little adjustment, one of the two bright lines can be brought to shine through the slit b, both of the lines being at the same time visible like a pair of stars at x, where they cross the slit a. Now, bring the incandescent lime behind the flame; the slit b will immediately increase considerably in brightness, but a will be many times brighter yet, and the two stars at x will be replaced by black dots apparently.

the solar spectrum those lines dark which are bright in the spectrum of sodium-vapor; and we do. If magnesium is there, it ought to manifest itself in the same way, and it does; and similarly for all the substances which spectrum analysis reveals.

If this view is correct, it follows also that this atmosphere, containing in gaseous form the substances whose presence is manifested by the dark lines of the ordinary spectrum—the sun's *reversing layer*, as it is now often called—would give a spectrum of bright lines if we could isolate its light from that of the photosphere. The observation is possible only under peculiar circumstances. At a total eclipse of the sun, at the moment when the advancing moon has just covered the sun's disk, the solar atmosphere of course projects somewhat at the point where the last ray of sunlight has disappeared. If the spectroscope be then adjusted with its slit tangent to the sun's image at the point of contact, a most beautiful phenomenon is seen. As the moon advances, making narrower and narrower the remaining sickle of the solar disk, the dark lines of the spectrum for the most part remain sensibly unchanged, though becoming somewhat more intense. A few, however, begin to fade out, and some even turn palely bright a minute or two before the totality begins. But the moment the sun is hidden, through the whole length of the spectrum, in the red, the green, the violet, the bright lines flash out by hundreds and thousands, almost startlingly; as suddenly as stars from a bursting rocket-head, and as evanescent, for the whole thing is over within two or three seconds. The layer seems to be only something under a thousand miles in thickness, and the moon's motion covers it very quickly.

The phenomenon, though looked for at the first

eclipses after solar spectroscopy began to be a science, was missed in 1868 and 1869, as the requisite adjustments are delicate, and was first actually observed only in 1870. Since then it has been more or less perfectly seen at every eclipse. Except at an eclipse it has not yet been found possible to observe this bright-line spectrum, because it is overpowered by the aërial illumination of our own atmosphere.

It is not, however, to be understood that the dark lines of the solar spectrum are due entirely or even principally to the stratum of gas which lies above the upper level of the photosphere. Were this so, the dark lines should be much stronger in the spectrum of light from the edges of the disk than in that from the center, which is not the case; at least, the difference is very slight. The photosphere, as we shall see hereafter, is probably composed of separate cloud-like masses floating in an atmosphere containing the vapors by whose condensation they are formed; the principal absorption, therefore, probably takes place in the interstices between the clouds, and below the general level of their upper limit.

The beautiful observations of Professor Hastings, of Baltimore, in which by an ingenious contrivance he managed to confront and compare directly the spectra of light from the center and edges of the sun's disk, have brought out the facts in the case very finely.

Theoretically, then, it is very easy to test the question of the presence of an element in the sun. It is only necessary to cover one half the length of the spectroscope-slit with a mirror or prism by which the sunlight is directed into the instrument, while at the same time a flame or electric spark, giving the spectrum of the substance under investigation, is placed directly in front of the other half of the slit. When matters are thus arranged, the observer sees in the instrument two spectra in juxtaposition, each of half the usual width— one the solar spectrum, the other that of the element under investigation; and it is easy to see whether the bright lines of the elementary vapor match exactly with corresponding dark lines in the solar spectrum.

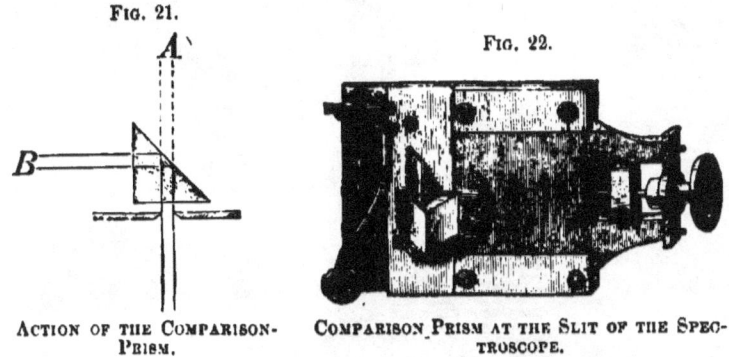

Fig. 21.

Fig. 22.

ACTION OF THE COMPARISON-PRISM.

COMPARISON PRISM AT THE SLIT OF THE SPECTROSCOPE.

The figures show the usual arrangement of the comparison-prism, as it is ordinarily called.

For the examination of the upper or violet portion of the spectrum, photography is employed with great advantage, the arrangement being precisely the same as that just indicated, except that a sensitized plate takes the place of the human retina, and the impression can be permanently retained for leisurely study. Certain light, too, as every one knows, which is invisible to the eye, strongly affects the photographic plate, so that the comparison can by this means be carried on into the ultra-violet and invisible regions of the spectrum.

The following full-page illustration is a representation of the arrangement of apparatus used by Mr. Lockyer in his celebrated researches—it is taken from his "Studies in Spectrum Analysis."

86 THE SUN.

Fig. 23.—General View of Spectrum Photographic Arrangements showing Heliostat Lamp and Lenses.

Theoretically, we say, the comparison is easy; but the practical difficulties are considerable. In the first place, it is not easy to get a spectrum of the body you wish to study, free from lines belonging to other substances—the requisite chemical purity is very troublesome to attain; and, in the next place, the dark lines of the solar spectrum are so numerous that it requires a very high dispersive power to establish a coincidence with certainty; a bright line in the spark-spectrum may fall *very* near a dark line with which it has no connection whatever. When, however, as in the case we have mentioned, the coincidences are not one or two, but numerous, and the lines in question peculiar in their character and appearance, a satisfactory result is soon established.

It was in this manner (by comparisons made by the eye and not by photography) that Kirchhoff in 1860 determined the presence in the solar atmosphere of the following elements: sodium, iron, calcium, magnesium, nickel, barium, copper, and zinc, the last two rather doubtful at that time. Since then the list has been greatly extended, and now stands as follows, according to the best authorities:

Elements.	Bright Lines in Spectrum.	Lines reversed in Solar Spectrum.	Observer.
1. Iron.............	600	460	Kirchhoff.
2. Titanium........	206	118	Thalen.
3. Calcium.........	89	75	Kirchhoff.
4. Manganese......	75	57	Ångstrom.
5. Nickel..........	51	33	Kirchhoff.
6. Cobalt..........	86	19	Thalen.
7. Chromium.......	71	18	Kirchhoff.
8. Barium..........	26	11	Kirchhoff.
9. Sodium.........	9	9	Kirchhoff.
10. Magnesium......	7	7	Kirchhoff.
11. Copper?........	15	7 ?	Kirchhoff.

Elements.	Bright Lines in Spectrum.	Lines reversed in Solar Spectrum.	Observer.
12. Hydrogen	5	5	Ångstrom.
13. Palladium †	29	5	Lockyer.
14. Vanadium †	54	4	Lockyer.
15. Molybdenum †	27	4	Lockyer.
16. Strontium *	74	4	Lockyer.
17. Lead	41	3	Lockyer.
18. Uranium †	21	3	Lockyer.
19. Aluminium †	14	2	Ångstrom.
20. Cerium *	64	2	Lockyer.
21. Cadmium	20	2	Lockyer.
22. Oxygen α } †	42	12 ± bright	H. Draper.
Oxygen β }	4	4 ?	Schuster.

The case of oxygen is peculiar, and will be considered more fully hereafter.

All of the above-named elements, except those marked with a †, are represented at times by bright lines in the spectrum of the chromosphere, which will be discussed in another chapter; and strontium and cerium were observed in that manner by the writer before the coincidence of their lines with *dark* lines in the ordinary solar spectrum had been satisfactorily made out. At least, two additional elements, as yet unidentified with any terrestrial substances, are recognized in the chromosphere by bright lines—one of them the unknown substance, which is most conspicuous in the corona, the other the hypothetical *helium*, as Frankland named it; there may, probably enough, be others also.

Besides the elements included in the above table, there is a certain probability that the following, viz., indium, lithium, rubidium, iridium, cæsium, bismuth, tin, silver, glucinum, lanthanum, yttrium, and carbon, are also present in the solar atmosphere, one or more lines of their spectra having been found by Mr. Lockyer to coincide with dark lines in the solar spectrum.

As to carbon, none of its characteristic lines appear in the visible portion of the solar spectrum; but, in the ultra-violet, Mr. Lockyer has discovered by photography a group of lines which are ascribed to this substance, so that its presence in the solar atmosphere is rendered probable.

On the other hand, the most careful observation fails to find, either in the ordinary spectrum or in that of the chromosphere, the slightest trace of silicon, chlorine, bromine, and iodine: of sulphur there are merely doubtful indications in the chromosphere spectrum. Some of Dr. Draper's photographs rather suggest, but only very uncertainly, the presence of nitrogen also.

When we recollect that the non-apparent elements constitute a great portion of the earth's crust, the question at once forces itself, What is the meaning of their seeming absence? Do they really not exist on the sun, or do they simply fail to show themselves; and, if so, why? The answer to the question is not easy, and astronomers are not agreed upon it. Mr. Lockyer has, however, proposed a theory which, if established, would remove most if not all of the spectroscopic difficulties. He thinks that our elements are not really elementary, but built of molecules themselves composite and capable of *dissociation* by the action of heat. Thus, a mass of chlorine, for instance, may at a certain temperature break up into constituents; and so it may easily be the case that at solar temperatures certain of our terrestrial elements can not exist; or, if they exist at all, can do so only in certain very restricted regions of the solar atmosphere.

One strong argument in favor of this view is found in the fact, now we think beyond dispute, that the same substance may, under different circumstances, give widely different spectra. Thus nitrogen and hydrogen each have two spectra, one a spectrum mostly composed of shaded bands, while the other consists of sharp, well-defined lines. Oxygen, according to Schuster's careful researches, has four spectra, and carbon is also assigned four by its investigators. There seem to be at least three possible explanations of these facts. One is, to suppose that the luminous substance, without any change in its own constitution, vibrates differently and emits different rays under varying circumstances, just as a metal plate emits various notes according to the manner in which it is held and struck. The second assumes that the substance, without losing its chemical identity, undergoes changes of molecular structure (assumes allotropic forms) under the varying circumstances which produce the changes in its spectrum. According to either of these views, although we can safely infer, from the presence of the known lines of an element in the solar spectrum, its presence in the solar atmosphere, we can not legitimately draw any negative conclusion: the substance may be present, but in such a state under the solar conditions as to give a spectrum different from any with which we are acquainted.

The other and simplest explanation is to suppose, with Mr. Lockyer, that the changes in the spectrum of a body are indications of its decomposition, the spectrum of the original substance being replaced by the superposed spectra of its constituents.

Another point which favors Mr. Lockyer's view is this: Certain substances have numerous lines apparently common. Thus, if one runs over Ångstrom's map of the solar spectrum, he will find about twenty-five lines marked as belonging both to iron and calcium. The same thing is true of iron and titanium to a still greater extent, and to a considerable degree of several other pairs of substances. This fact might be explained in several ways. The common lines may be due—first, to impurities in the materials worked with; or, second, to some common constituent in the substances (which is Lockyer's view); or, third, to some similarity of molecular mass or structure which determines an identical vibration-period for the two substances; or, finally, it may be that the supposed coincidence of the lines is only apparent and approximate—not real and exact—in which case a spectroscope of sufficient dispersive power would show the want of coincidence.

Now, Mr. Lockyer, by a series of most laborious researches, has proved that many of the coincidences shown on the map are merely due to impurities, and he has been able to point out which of the lines mapped as common to calcium and iron, for instance, belonged to each metal. As the iron employed is rendered successively purer and purer, certain of the common lines become fainter, and such evidently belong to calcium and not to iron. Similarly, when calcium is used, we can point out the lines which are due to the iron contamination. But, when all is done, we find that certain of the common lines persist, becoming more and more conspicuous with every added precaution taken to insure purity of materials.

Moreover, when one of the substances, say the calcium, is subjected to continually increasing temperatures, its spectrum is continually modified, and these *basic-lines*, as Mr. Lockyer calls them, are the ones which become increasingly conspicuous, while others disappear. This is just what ought to happen if they are due to some element common to both the iron and calcium—an element liberated in increasing abundance with every rise of temperature.

One who wishes to see the argument fully stated, must refer to Mr. Lockyer's own papers, published for the most part in the "Proceedings of the Royal Society" in 1878 and 1879. The case is certainly a very strong one, and the hypothesis would give a very simple account of the state of things upon the sun and in the stars. It assumes that they are merely too hot to permit the existence in their atmospheres of the missing substances, which, according to this view, dissociate or break up at lower temperatures. If this be so, we may perhaps some time, by the help of the electric arc or spark, be able to produce a similar result in our laboratories, and exhibit the components of oxygen, chlorine, or carbon. Indeed, some experiments of Meyer, of Zürich, in 1878, seem to show that chlorine is a compound containing oxygen, though a different explanation has been suggested.

On the other hand, the new doctrine is inhospitably received by many chemists, since it is very difficult to reconcile it with the laws which have been found to connect the chemical constitution and atomic weight of bodies.

In a considerable number of cases, also, the powerful spectroscopes of Thollon and others have shown these basic-lines to be close *doubles*, as for instance b_3, b_4, and E; so that, in these instances, we probably have to do with lines of different substances not actually coincident, but only accidentally near each other in the spectrum. The writer has recently made a very careful examination of the seventy lines shown on Ångström's map as common to two or more substances, using the powerful diffraction spectroscope described on a preceding page (Chapter III, pages 10, 11). Out of the whole number of lines, fifty-six are distinctly double or triple, seven appear to be single, and as to the remaining seven it is uncertain. Two of these uncertain lines are strongly suspected to be double, and the other five can not be identified with certainty, because they fall upon spaces thickly covered with groups of fine lines not shown in the map. In respect to three of the seven lines which appear to be single, there is a disagreement between Ångstrom's map and Thalen's tables which accompany the map. As matters stand at present, therefore, very little weight can be assigned to the argument depending upon the supposed coincidence of lines. If, however, it should hereafter turn out that any of these double lines appear as double in the spectra of *both* the elements to which Ångstrom and Thalen have assigned them, the argument will at once regain more force than it ever had before the resolution of the lines, and would be unanswerable, as a proof of some community of substance or structure in the molecules of the elements concerned.

But if we reject this hypothesis, that our so-called elements are really not elementary, it becomes a very troublesome matter to account for the non-appearance of the lines of the missing substances in the solar spectrum. Possibly, in some cases, the very brilliance of the lines of an element may prevent their appearance as dark lines. It is possible, for instance, to make the bright lines of sodium so intense that the light from an incandescent lime-cylinder will not be able to reverse them, and, of course, by making them a little less intense, they may be caused to disappear entirely, being neither brighter nor darker than the continuous spectrum on which they are projected. This actually seems to be the case with the hypothetical helium, which gives in the chromosphere spectrum an intensely brilliant yellow line, known as D_3, because it is very near to the sodium-lines, D and D_2. At times, and especially in the neighborhood of sun-spots, a very faint dark line marks its place, but usually the spectrum of the photosphere fails to give the slightest indication of its presence. There are fourteen similar cases in different parts of the spectrum, but only three or four of them are probably identifiable with the lines of any terrestrial element.

In this case, however, the element, whatever it may be, though not represented by a dark line in the spectrum of the photosphere, still is represented in another and an intelligible manner; but the missing elements make no sign whatever.

The case of oxygen is peculiar. It does not show itself by any conspicuous dark lines; neither are the bright lines of its ordinary spectrum found in the chromosphere. In 1877 Dr. Henry Draper, of New York, announced that he had discovered its presence in the sun, and he published photographs which show, in a very convincing manner, the coincidence between the bright lines of this element and certain *bright* spaces or bands in the solar spectrum. His method of procedure was to form the spectrum of oxygen by means of sparks from a powerful induction-coil, worked by a dynamo-electric machine, itself driven by an engine. These sparks passed between iron terminals, in a little chamber wrought out of soapstone, through which a current of pure oxygen was forced at atmospheric pressure nearly; sometimes, however, air was used instead, giving the same results, except that the spectrum of nitrogen was then superadded to that of oxygen. The spectrum of this spark was photographed simultaneously with that of the sun, the sunlight being brought in through half the slit by a small reflector, and thus a comparison was obtained, free from personal bias, between the solar spectrum and that of the gas. The iron lines, due to the terminals, are a great assistance in testing the adjustments. The oxygen lines produced in this way at atmospheric pressure are not so well defined as those seen in the spectrum of a Geissler tube, but are rather broad and hazy.

In the blue portion of the solar spectrum, which alone is accessible to photography, the Fraunhofer lines are generally very numerous, close, and black; but here and there is an interval free, or comparatively free, from lines. In a low-dispersion spectroscope such an interval looks like a bright band. Now, almost every one of the dozen or so bright lines of oxygen, which the photographs display, falls exactly against one of these brighter interspaces.

It is hardly possible that this can be merely due to chance; and a careful study of the photographs satisfies almost every one that in some way or other solar oxygen must be concerned in the phenomenon. Dr. Draper has since repeated these laborious and expensive experiments in a still more elaborate manner, and with results entirely confirmatory of those first reached.

It is, however, extremely difficult to explain how oxygen in the sun's atmosphere can produce such an effect in the ordinary solar spectrum while remaining invisible in the spectrum of the chromosphere; and the most careful search does not show a single one of *these* bright oxygen-lines. We say of *these* lines, because Dr. Schuster has shown, with great probability, that a different oxygen spectrum, with only four bright lines in it, has these four all represented by dark lines in the photospheric spectrum, and two of the four in the spectrum of the chromosphere.

It is only fair to those who still dissent from Dr. Henry Draper's opinion, as many eminent authorities do, to say that, with high dispersive powers, the "bright bands" of the solar spectrum entirely lose their prominence, and are even found to be occupied by numerous fine dark lines. Dr. John C. Draper has suggested that these dark lines may be the true representatives of oxygen.

It will be seen, of course, that Mr. Lockyer's view removes most of the difficulties, but not all. Unless Dr. H. Draper's photographs are entirely deceptive in their coincidences, we have yet something to learn as to the formation of spectra under solar conditions.

The lines of the solar spectrum not only inform us as to the presence or absence of bodies in the solar atmosphere, but give us, to some extent, indications as to their physical condition. The spectrum of a given body, say hydrogen, varies very much in the relative strength and brightness of its lines, according to the circumstances of its production. If, for instance, the gas be highly rarefied, and the electric spark, which illuminates it, not too strong, the lines will be fine and sharp. Under higher pressure and more intense discharges, some of them will become broad and hazy, and new lines, before unseen, will make their appearance. So of other substances; and this apart from the fact, before stated, that a given element often has several entirely different spectra. Changes, such as have been mentioned, go on up to a certain point, and then, suddenly, an entirely new spectrum appears, not having apparently the slightest connection with the one which preceded it any more than if it came from an entirely different element or mixture of elements; as, in fact, according to Mr. Lockyer's view, is probably the case.

Now, in the solar spectrum, the dark lines characteristic of an element are all coincident with bright lines of its gaseous spectrum; but it is not often the case that the relative width and intensity of the solar lines match those of the bright lines in the spectrum obtained by artificial means. In the spectrum of calcium, for instance, certain lines, which in our laboratory experiments are the most conspicuous, are very faint upon the sun, and others, which are inconspicuous in the spark spectrum, are vastly more important on the solar surface. As yet, we are not able with certainty to interpret all these variations, but, in a general way, it may be said that they all point to the conclusion that the temperature of the solar atmosphere is considerably higher than that of any of our flames or electric arcs or sparks.

At times, also, when the motions of the solar atmosphere become unusually intense, the spectroscope apprises us of the fact, and gives us the means of determining the rate at which the moving masses are advancing toward us or receding from us. If a luminous body is approaching with a velocity at all comparable with that of light, the *pitch* of the light, if the expression may be allowed—its wave-length and number of vibrations per second—will be changed and heightened just as in the case of sound.

Most of our readers have probably noticed the curious change in pitch of the bell or whistle of a locomotive passing at full speed, especially if we ourselves were on a train moving in the opposite direction. If the velocity is great, about forty miles an hour for each of the two trains, the pitch will drop a full major third.

The explanation is simply this: If both ourselves and the locomotive carrying the bell were at rest, we should hear the bell's true sound, the pulsations following each other at regular and the real intervals. If, now, we are rapidly approaching the bell, the interval of time between the impact of each pulse upon the ear and the following one will be shortened, because after any pulse has been received we advance part way to meet the next, and so encounter it earlier than if we had remained at rest. Now, this interval of time between successive pulsations is precisely what determines the pitch of the sound: the more pulsations there are in a second the higher the pitch. It is obvious that, if we remain at rest and the bell approaches us, the same effect will be produced, and that, if both are moving, the effects will be added; and, finally, it is clear that the recession of the hearer from the bell will produce the opposite effect and lower its pitch.

Just the same thing holds good of light; it also consists of pulsations, and the refrangibility of a ray and its *diffrangibility*, if we may coin the word, both depend upon the number of pulsations per second with which it reaches the diffracting or refracting surface. The more frequent the pulsations the more it will be refracted, and the less it will be diffracted. If, then, we were swiftly approaching a mass, say of incandescent hydrogen, we should find the position of each of its characteristic rays in the spectrum slightly altered, and falling farther from the red end of the spectrum (the region of slow vibrations) than if we were at rest. By comparing the positions of these lines with those obtained from a Geissler tube containing hydrogen, we could find how much change was produced, and therefore how the velocity with which we are approaching the moving mass compares with that of light. Similarly, if the body were advancing toward us. And, *vice versa*, if the distance were increasing, the lines would be shifted downward in the spectrum toward the red.

Because the velocity of light is exceedingly great (more than 186,000 miles per second), it is evident that only very swift motions can produce any sensible displacement of lines in the spectrum. Since, however,

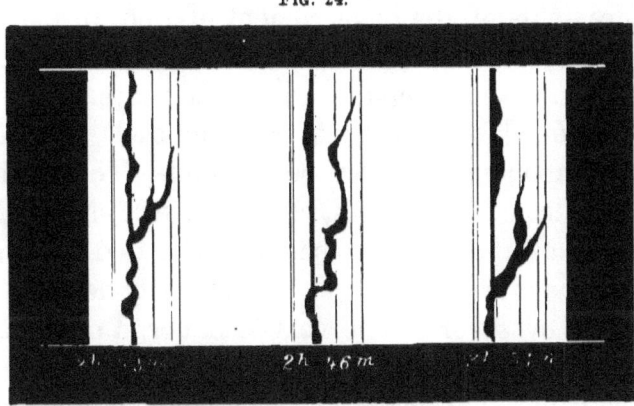

Fig. 24.

CHANGES IN THE C LINE (September 22, 1870).

in the neighborhood of sun-spots and in the solar prominences, we frequently meet with masses of gas moving from thirty to fifty miles a second, and sometimes as much as three hundred miles a second, it is not unusual, in working with the telespectroscope, to observe the distortion and displacement of portions of a dark line which are produced by these motions, and indicate them.

The figure represents the appearance of the C line seen in the spectrum of a sun-spot by the writer on September 22, 1870. The velocities indicated vary from two hundred and thirty to three hundred and twenty miles per second; the latter is seldom, if ever, exceeded.

Results of this sort are so surprising that there have been many attempts to escape from them, and to account for the distortion of lines in some other way, but without any satisfactory success. There have been difficulties raised also in regard to the mathematical theory of the matter. These have been met, however; and what amounts to an experimental verification of the correctness of the received view has been reached by measurements of the displacement of lines in the spectra of the eastern and western limbs of the sun. The eastern limb is moving toward us, the western from us, in consequence of the sun's rotation, each with a velocity of about 1·25 miles per second. The resulting displacement of the lines is, of course, very slight—only about $\frac{1}{100}$ of the distance between the two D lines— but, small as it is, it has been satisfactorily detected and measured by several observers—Zöllner, Vogel, Langley, and the writer, among them.

The values determined have ranged generally somewhat larger than 1·25. My own result was $1·42 \pm 0·07$. The difference is a little larger than the probable error seems to justify, and may very possibly indicate a physical fact: that the solar atmosphere is really drifting forward over the photosphere. But this needs confirmation before it can be accepted as certain.

In the motion-distortions of lines Lockyer finds strong confirmation of his ideas. It not unfrequently happens that in the neighborhood of a spot certain of the lines which we recognize as belonging to the spectrum of iron give evidence of violent motion, while close to them other lines, equally characteristic of the laboratory spectrum of iron, show no disturbance at all. If we admit that what we call the spectrum of iron is really formed in our experiments by the superposition of two or more spectra belonging to its constituents, and that on the sun these constituents are for the most part restricted to different regions of widely varying pressure, temperature, and elevation, it becomes easy to see how one set of the lines may be affected without the other.

The same facts are, of course, also explicable on the supposition that there are several allotropic forms of iron-vapor, mixed together in terrestrial experiments but separated on the sun, and sorted out, so to speak, by the conditions of temperature and pressure.

CHAPTER IV.

SUN-SPOTS AND THE SOLAR SURFACE.

Granulation of Solar Surface.—Views of Langley, Nasmith, Secchi, and others.—Faculæ.—Nature of the Photosphere.—Janssen's Photographs of Solar Surface—the *Reseau Photospherique.*—Discovery of Sun-spots.—General Appearance and Structure of a Spot.—Its Formation and Disappearance.—Duration of Sun-spots.—Remarkable Phenomena observed by Carrington and Hodgson.—Observations of Peters.—Dimensions of Spots.—Proof that Spots are Cavities.—Sun-spot Spectrum.—" Veiled Spots."—Rotation of Sun.—Equatorial Acceleration.—Explanations of the Acceleration.—Position of Sun's Axis and Secchi's Table for its Position Angle at Different Times of the Year.—Proper Motions of Spots.—Distribution of Spots.

WHEN an observer, provided with suitable telescopic appliances, examines the surface of the sun, he finds a most interesting field before him. At first view, indeed, it is less impressive than the moon; there is not so much to attract the immediate attention—no mountain-ranges and craters, no shadows, rills, or rays.

But, if the telescope is a good one and the atmospheric conditions favorable, the details soon begin to come out: the surface is seen to be far from uniform, composed of minute grains of intense brilliance and irregular form, floating in a darker medium, and arranged in streaks and groups. If the magnifying power employed is rather low, the general effect of the surface is much like that of rough drawing-paper, or of curdled milk seen from a little distance; and, generally speaking, a low power is all that can be used, because the heat of the sun commonly keeps the air in a state of great disturbance, so that it is only occasionally that the solar surface can be scrutinized with such powers as we continually employ upon the moon and planets. But now and then times come—favorable minutes, and even hours —when the telescopic power can be pushed to its maximum, and we get such views as that which Professor Langley has presented in the beautiful drawing of which our frontispiece is a reproduction. The grains, or "nodules," as Herschel called them, are then seen to be irregularly rounded masses, measuring some hundreds of miles each way, sprinkled upon a less brilliant background, and making much the same impression as snowflakes sparsely scattered over a grayish cloth, to use the comparison of Professor Langley. If the telescope has a diameter of not less than nine inches, and if the seeing is absolutely exquisite, then these grains themselves are sometimes resolved into "granules," little luminous dots not more than a hundred miles or so in diameter, which, by their aggregation, make up the grains, just as they in their turn make up the coarser masses of the solar surface. Professor Langley estimates these granules to constitute perhaps about one fifth of the surface of the sun, while they emit at least three quarters of the light. He and Secchi seem to be so far the only observers who have ever fairly seen them. The "grains" have been known for years and described by many observers, but with some very embarrassing discrepancies. Nasmyth, in 1861, described them as "willow-leaves" in shape, several thousand miles in length, but narrow, with pointed ends; and figured the surface of the sun as a sort of basket-work formed by the interweaving of such filaments. Fig. 25 is copied from one of his pictures. His statement excited a good deal of

104 THE SUN.

Fig. 25.

Group of Solar Spots observed and drawn by Nasmyth (June 5, 1864).

pretty warm discussion. Dawes entirely denied the existence of any such forms; while Stone and Secchi assigned them much smaller dimensions, and compared them to rice-grains. Huggins agreed completely with

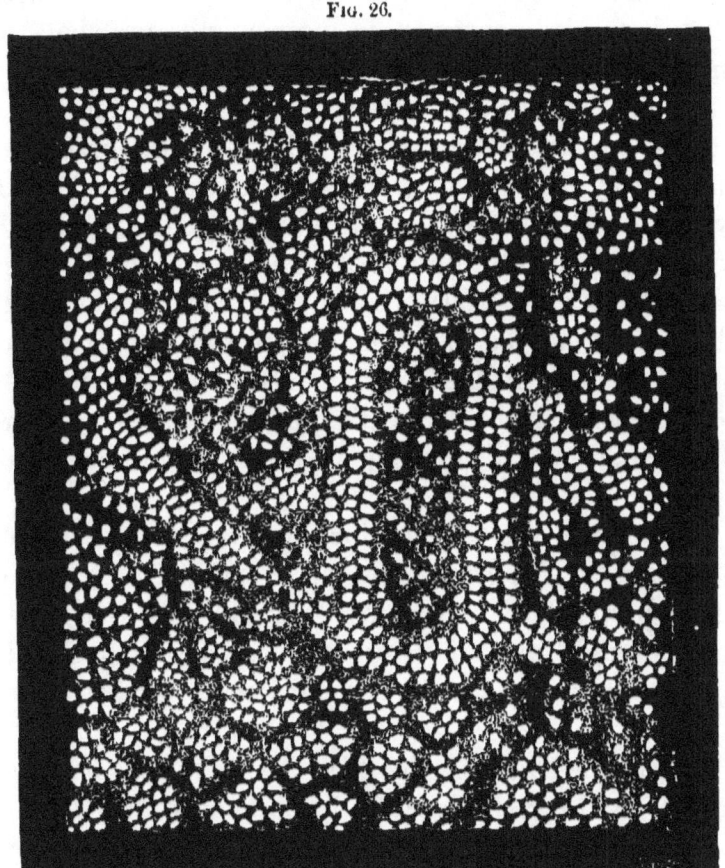

Fig. 26.

GRANULES AND PORES OF THE SUN'S SURFACE. (After Huggins.)

neither, but represents the "make-up" of the solar surface by a drawing from which Fig. 26 is taken. This is unquestionably a very correct delineation of what is seen with a good telescope under circumstances fair, but not the best possible.

On portions of the sun's disk, however, the elementary structure is often composed of long, narrow, blunt-ended filaments, not so much like "willow-leaves" as like bits of straw, lying roughly parallel to each other —a "thatch-straw" formation, as it has been called. This is specially common in the penumbræ of spots, or in their immediate neighborhood.

If one were to speculate as to the explanation of the grains and thatch-straws, it might be that the grains are the upper ends of long filaments of luminous cloud, which, over most of the sun's surface, stand approximately vertical, but in the penumbra of a spot are inclined so as to lie nearly horizontal. This is not certain, though; it may be that the cloud-masses over the more quiet portions of the solar surface are really, as they seem, nearly globular, while near the spots they are drawn out into filamentary forms by atmospheric currents.

Whatever the explanation may be, the appearance of things in the immediate neighborhood of a spot is often pretty fairly represented by Mr. Nasmyth's pictures, though that of Professor Langley is decidedly more accurate in details, and represents far better seeings.

Near the edges of the disk the light falls off very rapidly, and certain peculiar formations, called the faculæ, are there much more noticeable than near the center of the disk. These faculæ (Latin, "a little torch") are irregular streaks of greater brightness than the general surface, looking much like the flecks of foam which mark the surface of a stream below a waterfall. Not unfrequently they are from five to twenty thousand miles in length, covering areas immensely larger than any terrestrial continent.

The figure, taken from a photograph by De La Rue, gives a reasonably correct idea of the general appearance of these objects, and of the darkening at the limb of the sun. No woodcut, however, is quite competent to give the delicate flocculence of the details. These faculæ are elevated regions of the solar surface, ridges and crests of luminous matter, which rise above the general level and protrude through the denser por-

Fig. 27.

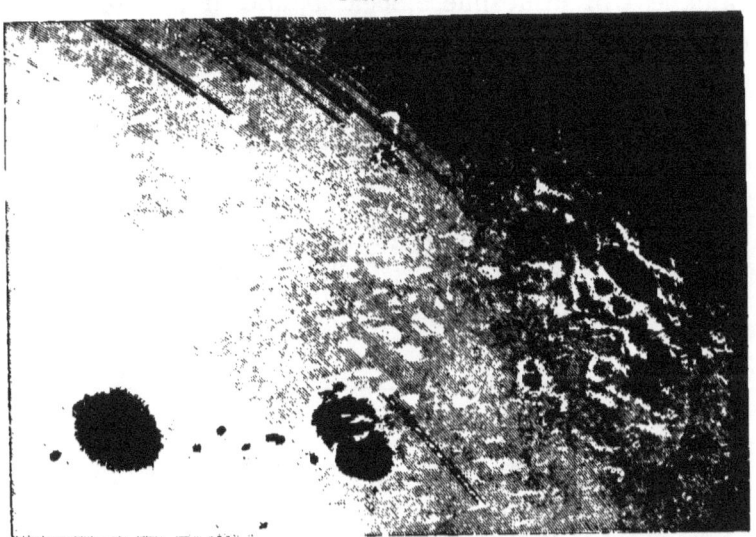

Sun-Spots and Faculæ. (From a Photograph.)

tions of the solar atmosphere, just as do our terrestrial mountains. The evidence of this is that, occasionally, when one of them passes over the edge of the disk, it can be seen to project like a little tooth—the reader should not forget, however, that the elevation, to be perceptible at all, must be at least one half a second of arc; that is, two hundred and twenty-five miles, or some forty-five times as high as any Himalaya.

The reason why they are so much more conspicuous

near the limb is simply this: The luminous surface is covered, as has been intimated before, with an atmosphere which is not very thick compared with the dimensions of the sun, but still sufficient to absorb a good deal of the light. Light coming from the center of the disk penetrates this atmosphere, as is apparent from the figure at *a*, under the most favorable conditions, and is but slightly reduced in amount. The edges of the disk, on the other hand, are seen through a much greater thickness of atmosphere, as at *b*, and are, therefore, of course, much obscured, the amount of absorption being

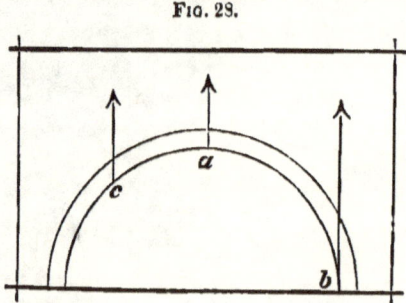

Fig. 29.

by some observers put as high as seventy five per cent. If, now, to take an extreme case, we suppose a facula high enough to lift its summit quite through this atmosphere, it will itself suffer no diminution of brilliance while the sun's rotation carries it from the center of the disk to the limb, but it will have passed from a background of brightness almost equal to its own, on which it would be seen only with difficulty, to another seventy-five per cent. or so darker, and will thus become very conspicuous. What is true of faculæ of such extreme dimensions is, of course, also measurably true of those of inferior elevation.

The faculæ are found to some extent over the whole

surface of the sun, though only sparingly in the polar regions, but they are especially abundant in the immediate neighborhood of spots, as the engraving well shows. In fact, it is almost as unusual to find a sun-spot without accompanying faculæ as a valley on the earth without surrounding hills. But the parallel is not quite exact, for there are numerous faculæ without neighboring spots.

Except near the spots, the faculæ change form and place, for the most part, rather slowly, persisting sometimes for several days without any very apparent alteration. Still, close observation and micrometric measurement will always detect some movement or deformation, even within an interval of only an hour or two; and near the spots the changes are often so rapid and extreme as to puzzle even a skilled draughtsman to keep up with them.

This, of course, shows that the faculæ are not to be compared with mountains; they are not permanent and stable, nor is the surface of the sun continental or oceanic even, but either a sheet of flame or of cloud rolling and tossing, and never at rest. When we come to study the minute details of the granulation, we find movements at the rate of a thousand miles an hour to be the rule rather than the exception.

And, although this is not the proper place to treat the subject at length, we may add that all we can learn as to the temperature and constitution of the sun makes it hardly less than certain that the visible surface, which is called the photosphere, is just a sheet of self-luminous cloud; precisely like the clouds of our own atmosphere, with the exception that the droplets of water which constitute terrestrial clouds are replaced on the sun by drops of molten metal, and that the solar atmosphere in which they float is the flame of a burning, fiery furnace, raging with a fury and an intensity beyond all human conception. Looking at it ninety-three million miles away, we fail at first to see, in such objects as faculæ and granules, the evidence of such commotion; but, when we convert our micrometric measurements of barely perceptible changes into miles and velocities, and figure to ourselves the scale of movement, we gradually comprehend their meaning, and begin to understand what we are dealing with.

A very great advance in our knowledge of the solar surface has recently been made through the photographic work of Janssen, mentioned in a previous chapter.* Many of his pictures (in which the disk of the sun measures about eighteen inches in diameter) show the details of the surface nearly if not quite as well as any visual observations; and with the advantage that, while the observer with the eye could only command a small field of view, one can, on these photographic plates, command the whole at once, and catch the relations of different parts. On examining one of these magnificent plates, one is at first struck with a sort of "smudginess" (to use the expression of Mr. Huggins in describing them), which might give the impression that it was not properly cleaned before coating with the collodion. A closer examination, however, shows that the peculiarity is not in the plate but in the image; there are patches of clear definition, half an inch or so in diameter upon a picture of the size mentioned, separated by streaks and patches where everything is indistinct and confused.

One might naturally attribute this to the disturbance of the air in the telescope-tube, and to clouds of vapor

* See page 59.

6 h. 47 m.

7 h. 37 m.

PHOTOGRAPHS OF A PORTION OF THE SUN,
BY JANSSEN,

Meudon, June 1, 1878. *Interval,* 50 *minutes.*

rising from the damp collodion surface when struck by the flash of sunlight during its exposure; but Janssen has found that pictures taken in immediate succession show the same "smudges" on the same parts of the sun, which, of course, would not happen if they were the result of accidental currents of air or vapor in the telescope-tube. He infers, therefore, that the phenomenon is solar, and has given it the name of the *Reseau Photospherique*, or "Photospheric Reticulation," since the streaks and patches of indistinctness cover the surface like a net.

The discovery of this feature in the structure of the solar surface is so far the most interesting and important result of astronomical photography.

While pictures taken in immediate succession exhibit the same details of reticulation, those taken at intervals of an hour or two show great changes, especially near spots and faculæ. We present on the opposite page a pair of such photographs, borrowed from the "Annuaire" of the Bureau of Longitudes for 1879. The original pictures were taken by Janssen, at Meudon, on June 1, 1878, with an interval of fifty minutes between them. They show clearly the peculiar characteristics of the *reseau photospherique*, as well as the nature and extent of the changes which occur in so short a time. Compare, especially, the granulation in the lower right-hand corner of each picture, and immediately around the upper spot, remembering all the while that the scale of the picture is about forty-six thousand miles to the inch, and that the little spot at the top of the figure is nearly seven thousand miles in diameter.

The idea of M. Janssen is that the regions of indistinctness are those where we look down upon the surface through a portion of the sun's atmosphere which is at the moment especially agitated, while the parts where the details of the granulation are clear and well defined are those which, at the moment, are covered by an atmosphere unusually quiet and homogeneous. These regions are continually interchanging with each other, just as areas of storm and fine weather sweep over the surface of the earth, but with inconceivably greater swiftness.

It is not, however, certain that the disturbed portions of the solar atmosphere, which produce the indistinctness in question, lie near the sun's surface. It may be that they are high up, and it would not be an unreasonable conjecture to suppose that the streamers and luminous masses of the corona may be concerned in the phenomenon; it is almost certain that any great aggregation of chromospheric matter would modify the appearance of whatever might be situated beneath it. The simple fact is, of course, that we are looking down upon the granules and other features of the sun's surface, not through an atmosphere shallow, cool, and quiet like the earth's, but through an envelope of matter, partly gaseous and partly, perhaps, pulverulent or smoke-like, many thousand miles in depth, and always most profoundly and violently agitated.

But, if there happens to be a well-formed group of spots upon the solar surface, they will be sure to claim the attention of one who, for the first time, looks at the sun through the telescope, quite to the exclusion of everything else. The umbra, with its central nuclei, and overlying bridges, veils, and clouds; the penumbra, with its delicate structure of filaments and plumes; the surrounding faculæ and the agitated surface of the photosphere in the whole neighborhood of the disturbance;

above all, the continual change and progress of phenomena—combine to make a fine sun-spot one of the most beautiful and intensely interesting of telescopic objects.

Even before the days of telescopes there are numerous records of dark spots seen by the naked eye upon the disk of the sun, especially in the annals of the Chinese. In the year 807 A. D., a large spot was visible in Europe for some eight days, and was supposed by many to be the planet Mercury, as was the case with a spot observed by Kepler in 1609; indeed, in all cases where such appearances were noted, they were invariably ascribed to bodies intervening between the earth and the sun. The idea of such imperfections upon the disk of a celestial body was utterly repugnant to the theological philosophy of the middle ages, and was admitted only slowly and grudgingly even after the demonstration of the fact was complete.

In 1610 and 1611 the discovery seems to have been made independently by Fabricius, Scheiner, and Galileo—Fabricius, according to our modern rules of scientific priority, being entitled to the credit as the first to publish the fact in a work, "De Maculis in Sole Observatis," which appeared at Wittenberg in June, 1611. The discovery was, of course, a necessary corollary to the invention of the telescope, which first came into use in Holland in 1608 or 1609. Fabricius's first observation was made in December, 1610. Galileo, in a letter responding to the account of Scheiner's discovery, and published early in 1612, claims to have seen the sun-spots with his newly-constructed telescope as early as October, 1610. Scheiner appears to have first seen sun-spots at Ingolstadt in March, 1611; but his ecclesiastical superior warned him against believing his own eyes in opposition to the authority of Aristotle, and it was not until November and December that he published an account of the matter in three letters to one Welser, a burgomaster of Augsburg, some months after the work of Fabricius had been printed. There is no reason whatever to doubt the word of Galileo, and his experience in losing the credit of this discovery, in consequence of his slowness of publication, seems to have been the origin of his curious method of publishing his subsequent discoveries in the form of anagrams, the interpretation of which was withheld for a time.

At the very outset of his observations, Fabricius, as well as Galileo, recognized that the spots are objects upon the surface of the sun, and that this body rotates on its axis, carrying them with it. Scheiner at first maintained that they were planets moving very near the sun, but not in contact with it. Many shared this opinion, and Tardé, a French astronomer, went so far as to name them the Bourbonian stars, in honor of the Bourbon dynasty. Scheiner's further observations soon convinced him, however, of the correctness of Galileo's opinion and arguments. Some twenty years later Scheiner published an enormous volume, the "Rosa Ursina," containing an account of his observations and apparatus. His telescope was mounted equatorially, and arranged to throw the sun's image upon a screen in precisely the manner employed by some of the best modern observers. He determined the time of the sun's rotation and the position of his equator with a very creditable degree of accuracy.

Since then observations upon these objects have been more or less kept up all the time, but not with any regular assiduity until within the last thirty years. It was soon found that they are only transitory and cloud-like in their nature, and interest in them therefore flagged, until their relations to the constitution of the sun began to be recognized.

A well-formed solar spot consists, generally speaking, of two portions—a very dark, irregular, central portion called the umbra, surrounded by a shade or fringe called the penumbra, less dark, and for the most part made up of filaments directed radially inward. The appearance of things, under ordinary circumstances of seeing, is as if the umbra were a hole, and the penumbral filaments overhung and partly shaded it from our view, like bushes at the mouth of a cavern. I say *as if*, and very possibly this is the actual case, the cen-

FIG. 29.

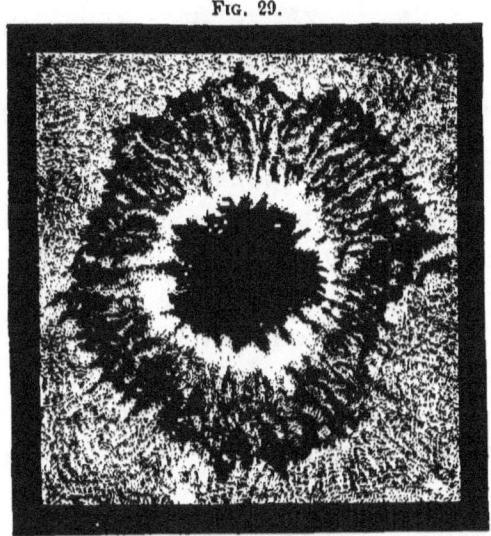

SPOT OF JULY 16, 1866.

tral portion being a real cavity filled with less luminous matter, and depressed below the general level of the photosphere, while the penumbra overhangs the edge.

The figure, copied from Secchi, is a fair representation of such a spot, and may be compared with the photographs of Janssen, which exhibit pretty much the same peculiarities, though with less of minute detail. The drawings of Nasmyth and Langley * show so much more of the detail than is ordinarily seen, that they are really less satisfactory representations of what one may expect when he observes a spot for the first time. Several points at once strike the attention. In the first place, the nearly circular form of the spot, which is the ordinary form during the middle life of one of these objects. While forming, and when on the point of disappearing, it is usually much more irregular. It is to be noticed also that there is nothing like a gradual shading off, either between the umbra and the penumbra or between the penumbra and the surrounding portions of the photosphere; on the contrary, the line of separation is strongly marked in each case, the penumbra being much brighter at the inner edge, and darker at the outer, so that it contrasts distinctly both with the umbra and with the neighboring surface of the sun. This brightness of the inner penumbra seems to be due to the crowding together of the penumbral filaments where they overhang the umbra. Again, it is observable that there is a general antithesis between the irregularities of the contour of the outer and inner edges of the penumbra. For the most part, where an angle of the penumbral matter crowds in upon the umbra, it is matched by a corresponding outward extension into the photosphere, and *vice versa*. It is noticeable also that many of the penumbral filaments are terminated by little detached grains of luminous matter, and there are also fainter veils of a substance less brilliant, but sometimes rose-colored, which seem to float above the umbra. Otherwise the umbra in the figure appears to be

* See frontispiece, and page 104.

uniformly dark;* but, if we had been actually observing the object on the 16th of July, 1866, when this picture was made, we should have found even the umbra full of detail—made up of cloudy masses of a brilliance really intense, and dark only by contrast with the still intenser brightness of the solar surface, as becomes apparent when the light from other portions is excluded. Probably we should have been able also to detect among these clouds one or more of the minute circular spots, first discovered by Dawes, much darker than the rest of the umbra, and presumably the mouths of tubular orifices penetrating to unknown depths.

If we were able to continue our watch for some time, we should see the details continually changing. The faint veils of overlying cirrus would probably melt away, and be replaced by others in some different position; the bright granules at the tips of the penumbral filaments would seem to sink and dissolve, and fresh portions would break off to replace them. We should find a continual indraught of the luminous matter over the whole extent of the penumbra. Almost certainly the spot would change its form and size, quite perceptibly from day to day, and sometimes even from hour to hour. Of course, we should find it steadily moving over the solar disk from the east toward the west, and as it neared the edge it would become apparently ellip-

---

* The umbra appears not black, but of a deep purplish tint. It is questionable, however, whether this color is real, or only due to the secondary spectrum of the telescope object-glass. The principal reason for suspecting this to be the case is in the fact that, during the transit of Mercury, in 1878, the planet's disk was found to present precisely the same tint, while there is no imaginable explanation for its really being anything but black. It is certain, too, on optical grounds, that any ordinary object-glass *must* show a purplish fringe extending inward over any dark spot upon a white background.

tical in form; the penumbra on the edge of the spot nearest the center of the sun would grow narrower and, perhaps, disappear entirely, and at last the spot, appearing like a mere line of darkness, but probably accompanied by an attendant crowd of faculæ, would pass out of sight behind the limb, perhaps to reappear again after a fortnight at the eastern edge. I say perhaps, because, quite as often as not, these short-lived objects are seen but once, not lasting through even a single revolution of the sun.

The average life of a sun-spot may be taken as two or three months; the longest yet on record is that of a spot observed in 1840 and 1841, which lasted eighteen months. There are cases, however, where the disappearance of a spot is very soon followed by the appearance of another at the same point, and sometimes this alternate disappearance and reappearance is several times repeated. While some spots are thus long-lived, others, however, endure only for a day or two, and sometimes only for a few hours.

The spots usually appear not singly, but in groups—at least, isolated spots of any size are less common than groups. Very often a large spot is followed upon the eastern side by a train of smaller ones; many of which, in such a case, are apt to be very imperfect in structure, sometimes showing no umbra at all, often having a penumbra only upon one side, and usually irregular in form. It is noticeable, also, that in such cases, when any considerable change of form or structure shows itself in the principal spot of a group, it seems to rush forward (westward) upon the solar surface, leaving its attendants trailing behind. When a large spot divides into two or more, as often happens, the parts usually seem to repel each other and fly asunder with great velocity—great, that is, if reckoned in miles per hour, though, of course, to a telescopic observer the motion is very slow, since one can only barely see upon the sun's surface a change of place amounting to two hundred miles, even with a very high magnifying power. Velocities of three or four hundred miles an hour are usual, and velocities of one thousand miles, and even more, are by no means exceptional.

At times, though very rarely, a different phenomenon of the most surprising and startling character appears in connection with these objects: patches of intense brightness suddenly break out, remaining visible for a few minutes, moving, while they last, with velocities as great as one hundred miles *a second*.

One of these events has become classical. It occurred on the forenoon (Greenwich time) of September 1, 1859, and was independently witnessed by two well-known and reliable observers, Mr. Carrington and Mr. Hodgson, whose accounts of the matter may be found in the monthly notices of the Royal Astronomical Society for November, 1859. Mr. Carrington at the time was making his usual daily observation upon the position, configuration, and size of the spots by means of an image of the solar disk upon a screen, being then engaged upon that eight years' series of observations which lies at the foundation of so much of our present solar science. Mr. Hodgson, at a distance of many miles, was at the same time sketching details of sun-spot structure by means of a solar eye-piece and shade-glass. They simultaneously saw two luminous objects, shaped something like two new moons, each about eight thousand miles in length and two thousand wide, at a distance of some twelve thousand miles from each other. These burst suddenly into sight at the edge of a great sun-spot, with a dazzling brightness at least five or six times that of the neighboring portions of the photosphere, and moved eastward over the spot in parallel lines, growing smaller and fainter, until in about five minutes they disappeared, after traversing a course of nearly thirty-six thousand miles. Their passage did not seem in any way to change the configuration of the spot over which they flew. Mr. Carrington found his drawing, which was completed just before they appeared, still quite correct after they had vanished. Of course, it is possible to question the connection between this phenomenon and the spot near which it appeared; but, as somewhat similar appearances have been seen by other observers since then, and always in the neighborhood of spots, it is probable that there is some relation in the case. Opinions have differed widely as to the explanation. Some have maintained that the phenomenon was simply due to the fall of a couple of immense meteors into the sun's atmosphere, others that it was caused by some sudden and powerful eruption from beneath, such as the spectroscope often reveals to us nowadays; an eruption, however, of most unusual brilliance and violence, for not one of the outbursts since then observed by the spectroscope has ever been visible without its aid.

A great magnetic storm and brilliant aurora followed this event that very night, and quite possibly were, in some way, caused by it—of which, more hereafter.

There is no regular process for the formation of a spot. Sometimes it is gradual, requiring days or even weeks for its full development, and sometimes a single day suffices. Generally, for some time before the appearance of the spot, there is an evident disturbance of the solar surface, manifested especially by the presence

SUN-SPOTS AND THE SOLAR SURFACE.

of numerous and brilliant faculæ, among which, "pores" or minute black dots are scattered. These enlarge, and between them appear grayish patches, in which the photospheric structure is unusually evident, as if they were caused by a dark mass lying veiled below a thin layer of luminous filaments. The veil grows gradually thinner, apparently, and breaks open, giving us at last the completed spot with its perfect penumbra. The "pores," some of them, coalesce with the principal spot, some disappear, and others constitute the attendant train before referred to. When the spot is once completely formed, it assumes usually an approximately circular form, and remains without striking change until the time of its dissolution. As its end approaches, the surrounding photosphere seems to crowd in upon and cover and overwhelm the penumbra. Bridges of light, often many times brighter than the average of the solar surface, push across the umbra, the arrangement of the penumbra filaments becomes confused, and, as Secchi expresses it, the luminous matter of the photosphere seems to tumble pell-mell into the chasm, which disappears and leaves a disturbed surface marked with faculæ, which in their turn subside after a time. As intimated before, however, the disturbance is not unfrequently renewed at the same point after a few days, and a fresh spot appears just where the old one was overwhelmed.

We transcribe from a paper by Dr. Peters, of Hamilton College, a very graphic account of the appearance and decay of certain sun-spots, based upon his observations at Naples in 1845–'46. It is printed in Volume IX of the "Proceedings of the American Association for the Advancement of Science." He says:

"The spots arise from insensible points, so that the exact moment of their origin can not be stated; but they grow very rapidly in the beginning, and almost always in less than a day they arrive at their maximum of size. Then they are stationary, I would say in the vigorous epoch of their life, with a well-defined penumbra of regular and rather simple shape. So they sustain themselves for ten, twenty, and some even for fifty days. Then the notches in the margin, which, with a high magnifying power, always appear somewhat serrate, grow deeper, to such a degree that the penumbra in some parts becomes interrupted by straight and narrow luminous tracks—already the period of decadence is approaching. This begins with the following highly interesting phenomenon: Two of the notches from opposite sides step forward into the area, over-roofing even a part of the nucleus; and suddenly from their prominent points flashes go out, meeting each other on their way, hanging together for a moment, then breaking off and receding to their points of starting. Soon this electric play begins anew and continues for a few minutes, ending finally with the connection of the two notches, thus establishing a bridge, and dividing the spot in two parts. Only once I had the fortune to witness the occurrence between *three* advanced points. Here, from the point A a flash proceeded toward B, which sent forth a ray to meet the former when this had arrived very near. Soon this seemed saturated, and was suddenly repelled; however, it did not retire, but bent with a sudden swing toward C; then again, in the same manner, as by repulsion and attraction, it returned to B; and, after having thus oscillated for several times, A adhered at last permanently to B. The flashes proceeded with great speed, but not so that the eye might not follow them distinctly. By an estimation of time and the known dimension of space traversed, at least an *under* limit of the velocity may be found; thus, I compute this velocity to be not less than two hundred millions of metres (or about one hundred and twenty thousand miles) in a second (*sic*).

"The process described is accomplished in the higher photosphere, and seems not to affect at all the lower or dark atmosphere. With it a second, or rather a third, period in the spot's life has begun, that of dissolution, which lasts sometimes for ten or twenty days, during which time the components are again subdivided, while the other parts of the luminous margin, too, are pressing, diminishing, and finally overcasting the whole, thus ending the ephemeral existence of the spot.

"Rather a good chance is required for observing the remarkable phenomenon which introduces the covering process, since it is achieved in a few minutes, and it demands, moreover, a perfectly calm atmosphere, in order not to be confounded with a kind of scintillation which is perceived very often in the spots, especially with fatigued eyes. The observer ought to watch for it under otherwise favorable circumstances when a large and ten- or twenty-days'-old spot begins to show strong indentations on the margin."

Dr. Peters, so far as we know, is the only observer who describes the remarkable phenomenon of *flashes* extending across an umbra with electrical velocity; and for this reason, and because his instrument was not of the highest power—a three-and-a-half-inch refractor—perhaps his account must be received with a little reserve until further confirmed. At the same time, there is nothing in the nature of the sun or of a sun-spot, so far as at present known, to make the statement in itself improbable; and certainly Dr. Peters holds deservedly a very high rank among astronomers for acuteness and accuracy of observation and description.

It must not be understood that the life-history of a spot, just sketched, applies to all, or even with exactness to a majority, of them. Almost every one has its own idiosyncrasies, departing in some respect or other from the usual course of things. Spots of unusual magnitude and activity often seem to have no quiet middle life; there is no time in their history when they are not doing something or other surprising, and more or less unprecedented.

We have spoken of the filaments which compose the penumbra as directed inward toward the center of the spot. This is the general rule, but the exceptions are numerous, and nothing can show better than Professor Langley's exquisite drawing how wide the divergence often is from this law. While at the left-hand and upper portions of the great spot (which, though "typical," is not a specimen of a quiescent spot) the filaments present the ordinary appearance, at the lower edge and upon the great overhanging branch they are arranged very differently. Very curious, and *rare*, also, though we have ourselves seen a similar thing on two or three occasions, is the feathery brush which reaches in below the "branch," so closely resembling a frost-crystal upon the window-pane in a winter's morning. What may be the cause of such formations it is now quite impossible to say. Probably analogies drawn from our terrestrial clouds will go further toward an explanation than any others yet proposed.

Not unfrequently the penumbral filaments are curved and spirally arranged, showing a marked cyclonic action. In such cases the whole spot usually turns slowly around, sometimes completing an entire revolution in a few days. More frequently, however, the spiral motion persists but a short time, and occasionally, after continuing for a while in one direction, the motion is reversed. Very often, in spots of considerable extent, there will be opposite spiral movements in different portions of the umbra; indeed, this is rather the rule than the exception. Neighboring spots show no tendency to rotate in the same direction. The number of spots in which a decided cyclonic motion appears is relatively quite small, not exceeding, according to the observations of Carrington and Secchi, more than two or three per cent. of the whole. Of course, these facts are sufficient to show that this kind of motion, when it occurs, is not attributable to anything like that action of the terrestrial atmosphere which determines the right- and left-handed rotation of our great storms in the southern and northern hemispheres. It is probably caused in sun-spots by merely accidental circumstances which convert the penumbral indraught into a whirl of no great rapidity or certain direction. It does not seem possible to find in this occasional cyclonic motion, as Faye attempts to do, the key and explanation of the whole series of sun-spot phenomena.

The dimensions of sun-spots are sometimes enormous. Many groups have been observed covering areas of more than one hundred thousand miles square, and single spots have been known to measure forty or fifty thousand miles in diameter, the central umbra alone being twenty-five or thirty thousand miles across. A spot, however, measuring thirty thousand miles over all, would be considered large rather than small.

An object of this size upon the sun's surface can easily be seen without a telescope when the brightness is reduced either by clouds, or nearness to the horizon, or by the use of a shade-glass. At the transit of Venus, in 1874, every one saw the planet readily without telescopic aid. Her apparent diameter was about 67″ at the time, which is equivalent to about thirty-one thousand miles on the solar surface. Probably a very keen eye would detect a spot measuring not more than twenty-three or twenty-four thousand miles.

Hardly a year passes, at times when spots are numerous, without furnishing several as large as this; so that it is rather surprising than otherwise that we have not a greater number of sun-spot records in the pre-telescopic centuries. The explanation probably lies in two things: the sun is too bright to be often or easily looked at, and when spots were seen they would be likely to be taken for optical illusions rather than realities.

During the years 1871 and 1872 spots were visible

to the naked eye for a considerable portion of the time. On several occasions pupils of the writer noticed them of their own accord, without having had their attention previously directed to the matter.

The largest spot yet recorded was observed in 1858. It had a breadth of more than one hundred and forty-three thousand miles, or nearly eighteen times the diameter of the earth, and covered about one thirty-sixth of the whole surface of the sun.

It has been intimated that the spots are depressions below the general level of the solar surface. The proofs are numerous, and the conclusion apparently unavoidable, though not without difficulties.

The fact was first clearly brought out by Dr. Wilson, of Glasgow, in 1769, and his demonstration was based upon the behavior of the penumbra of a spot which he observed in November of that year. He found that, when the spot appeared at the eastern limb or edge of the sun, just moving into sight, the penumbra was well marked on the side of the spot nearest to the edge of the disk, while on the other edge of the spot, that next the center, there was no penumbra visible at all, and the umbra itself was almost hidden, as if behind a bank. When the spot had moved a day's journey farther inward toward the center of the disk, the whole of the umbra came into sight, and the penumbra on the inner edge of the spot began to be visible as a narrow line. After the spot was well advanced upon the disk, the penumbra was of the same width all around the spot; but, when the spot approached the sun's western limb, the same phenomena were repeated as at the eastern—that is, the penumbra on the *inner* edge of the spot narrowed much faster than that on the outer, disappeared entirely, and finally seemed to hide from sight much of the umbra, nearly a whole day before the spot passed from view around the limb. Of course, it is hardly necessary to point out what the figure at once makes evident, that this is precisely the way things would go if the spot were a saucer-shaped de-

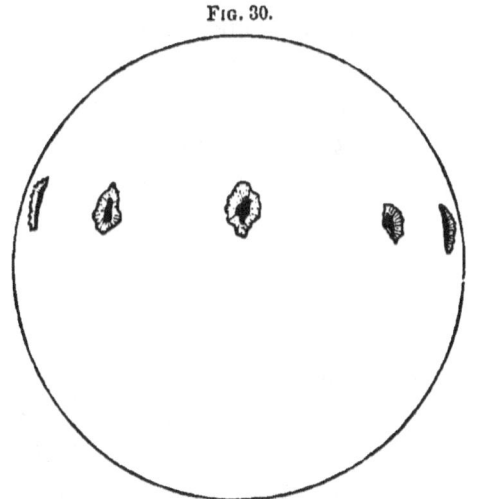

Fig. 30.

Diagram illustrating the Fact that Sun-Spots are Hollows in the Photosphere.

pression in the sun's surface, the bottom of the saucer corresponding to the umbra, and the sloping sides to the penumbra.

The observation of a single spot would hardly settle the question, because we frequently have spots with a one-sided penumbra. In fact, when spots are either in the process of formation or of dissolution the penumbra is seldom of uniform width all around. De La Rue, Stewart, and Loewy made, therefore, a few years ago, a careful discussion of something more than six hundred cases of spots, with measurable penumbræ, and found that, in a little over seventy-five per cent. of all the cases, the penumbra was widest on the edge of the spot nearest the limb, as Wilson's theory requires; in a little more than twelve per cent. there was no noticeable difference; and in the remaining twelve per cent. it was widest on the inner edge.

Others, Secchi especially, have investigated the matter by carefully measuring, from day to day, the position on the sun's disk of some selected point in the umbra of a spot. The work is not easy, and rather unsatisfactory, on account of the rapid changes, which make it difficult to identify the point of reference in successive observations; still, the result is quite decisive, showing, as an ordinary rule, that what may be called the "floor" of the umbra is depressed from two to six thousand miles, and sometimes more, below the general level of the photosphere.

On a few occasions, when a spot of unusual size and depth passes over the limb of the sun, a distinct depression is observed in the outline. Cassini describes such an instance in 1719. Herschel, De La Rue, Secchi, and others have given us several other observations of the same kind. Usually, however, the faculæ, which surround the spot, mask this effect entirely, and often actually give us a number of little projecting hillocks in place of the expected depression.

The spectrum of a sun-spot also furnishes an argument in the same direction, tending to show that the dark portion is a cavity filled with gases and vapors, which produce the obscuration, in part, at least, by absorbing the light emitted from the floor of the depression. It is not difficult to set the instrument in such a manner that the image of a sun-spot shall fall precisely upon the slit of the spectroscope. In this case the spectrum will be seen to be traversed by a longitudinal dark stripe, which is the spectrum of the umbra of the spot:

on each side is the spectrum of the penumbra, which is usually only a trifle fainter than that of the general surface of the sun. The width of the stripe, of course, depends upon the diameter of the spot. Along the whole length of the spot-spectrum the background is darkened, showing a general absorption; and in the upper part of the spectrum, from F to H, this seems to be pretty much all that can be noticed. In the lower part of the spectrum, however, from F downward, and especially between C and D, the spot-spectrum is full of interesting details and peculiarities, which deserve a far more

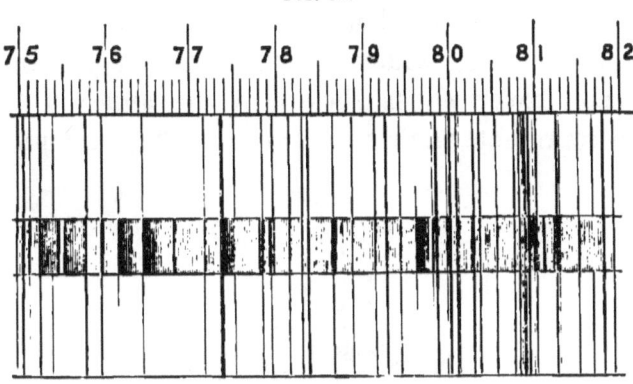

Fig. 31.

Portion of Sun-Spot Spectrum between C and D.

thorough and prolonged study than they have yet received. Many of the dark lines of the ordinary spectrum are wholly unmodified in the spectrum of the spot; in fact, this seems to be the case with the majority of them. Others, however, are much widened and darkened, and some, which are hardly visible at all in the ordinary spectrum, are so strong and black, in the penumbra even, as to be very conspicuous. Certain other lines, which are strong in the ordinary spectrum, thin out and almost disappear in the spot-spectrum, and some are even reversed at times. There are also a number of *bright* lines, not very brilliant, to be sure, but still unmistakable, and there are some dark shadings of peculiar appearance.

The annexed figure (Fig. 31), which represents a small portion of the spectrum of a spot observed by the writer in 1872, shows nearly all of these peculiarities. The portion represented lies between C and D, the scale attached being that of Kirchhoff's map.

Speaking in a general way, the lines of hydrogen, iron, titanium, calcium, and sodium, are more affected than those of other elements. The hydrogen lines are very often reversed; those of iron, titanium, and calcium, are usually thickened, and those of sodium are often enormously widened, and occasionally both widened and reversed, as shown in the annexed figure, which represents their appearance in the spectrum of a

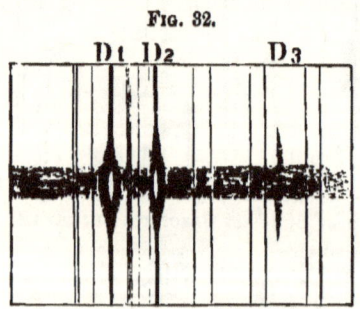

FIG. 32.

REVERSAL OF THE D-LINES.

spot observed on September 22, 1870. It will be noticed that at the same time the helium-line, D_3, which usually is invisible on the solar surface, was quite conspicuous as a dark shade. On this occasion the lines of magnesium also behaved in the same manner as those of sodium.

At times, also, the spectrum of a spot gives evidence of violent motion in the overlying gases by distortion and displacement of the lines. When the phenomenon occurs, it is more usually at points near the outer edge of the penumbra than over the central portion of the spot; but, occasionally, the whole neighborhood is violently agitated. In such cases it often happens that lines in the spectrum side by side are affected in entirely different ways—one will be greatly displaced, while its neighbor is not disturbed in the least, showing that the vapors which produce the lines are at different levels in the solar atmosphere, and do not participate to any great extent in each other's movements.

It is, perhaps, worthy of special remark that the two H lines appear to be reversed in the spectrum of a spot as a general rule—much more frequently, certainly, than any other lines known.

These lines, until very lately, have been ascribed by most authorities to calcium, but the recent investigations of Huggins, Vogel, Lockyer, and others, go to show that H_1, at least, should be assigned to hydrogen.

In a few instances the gaseous eruptions in the neighborhood of a spot are so powerful and brilliant that, with the spectroscope, their forms can be made out on the background of the solar surface in the same way that the prominences are seen at the edge of the sun. In fact, there is probably no difference at all in the phenomena, except that only prominences of most unusual brightness can thus be detected on the solar surface. An occurrence of this kind fell under the writer's observation on September 28, 1870. A large spot showed in the spectrum of its umbra all the lines of hydrogen, magnesium, sodium, and some others, reversed. Suddenly the hydrogen lines grew greatly brighter, so that, on opening the slit, two immense luminous clouds could be made out, one of them nearly 130,000 miles in length, by some 20,000 in width, the other about half as long. They seemed to issue at one extremity from two points near the edge of the penumbra of the spot. After remaining visible about twenty minutes, they faded gradually away, without apparent motion.

In addition to spots, such as we have been dealing with, there are occasionally seen on the solar surface dark-gray patches, which Trouvelot, who first called attention to them in 1875, has named "veiled spots," * considering that they are essentially of the same nature as other spots, but differing in this, that the disturbance which generates them is not sufficiently powerful to reach the surface and break entirely through the photosphere. Over these *veiled* spots the bright granules are less numerous and smaller than elsewhere, but much more mobile; sometimes, and frequently indeed, they are overlaid by faculæ. The changes of form and appearance in these objects are very rapid, affairs of a minute or two only, according to Trouvelot. They are found all over the solar surface, not being at all confined to the regions occupied by the ordinary spots, but sometimes occurring within eight or ten degrees of the sun's pole. They have been little observed, however, and information respecting them is as yet very meager.

ROTATION OF SUN AND PROPER MOTIONS OF SPOTS.

We have already mentioned that the spots travel across the disk of the sun, from the eastern edge to the western, in such a manner as to show that they are

---

* For Trouvelot's account of them, see "American Journal of Science and Art," March, 1876, Third Series, vol. xi.

attached to the surface, and that the sun rotates upon its axis. The true period is about twenty-five days, the apparent period being some two days longer, because the earth itself is continually moving forward in its orbit.

When we come, however, to study the motions of the spots more carefully, we find that they have movements of their own (*proper motions*, as astronomers call them), both in latitude and longitude, so that no observations of any single spot, however carefully conducted, can furnish an accurate determination of the position of the sun's axis and its period of rotation. This fact does not seem to have been comprehended by the early observers (though a neglected remark of Scheiner's indicates that he had a glimpse of the truth), and hence we have serious discordances between their different results, which range from 25·01 days, the result obtained by Delambre in 1775, to 25·58 days, as determined by Cassini about a hundred years earlier. The different values for the inclination of the sun's equator to the ecliptic lie between $6\frac{1}{2}°$ and $7\frac{1}{2}°$, and those for the longitude of the node between 70° and 80°. The most reliable recent results are those of Carrington and Spoerer. The former makes the *mean* period of the sun's rotation 25·38 days, while Spoerer gives it as 25·23.

The researches of Carrington,* between 1853 and 1861, first brought out clearly the fact that, strictly

---

* A memoir by Laugier, presented to the French Academy in 1844, but never published *in extenso*, contains, according to Faye, data which would lead to the same result. The summary, given in the "Comptes Rendus," fails, however, to indicate any appreciation of the *systematic* variation of rotation rate from equator to poles, and in no way invalidates Carrington's claim to be considered the discoverer of the law.

speaking, the sun, as a whole, has no single period of rotation, but different portions of its surface perform their revolutions in different times. The equatorial regions not only move more rapidly in miles per hour than the rest of the solar surface, but they *complete the entire rotation in shorter time.* If we deduce the period by means of spots near the sun's equator, we shall find it to be very nearly 25 days—a trifle less, according to Carrington. Spots at a solar latitude of 20° have, on the other hand, a period nearly 18 hours longer; at 30° the period rises to $26\frac{1}{2}$ days, and at 45° to $27\frac{1}{4}$, though in this latitude there are so few spots that the determination is not very reliable. Beyond this latitude we have nothing satisfactory, and it is not possible to determine, with any certainty, whether this retardation continues to the pole or not.

It is a curious circumstance, probably connected with this remarkable law of surface-movement, that the spots mostly lie between ten and thirty-five degrees of latitude on each side of the sun's equator; and it is this fact which makes it difficult to ascertain the exact laws of the solar rotation, since our observations are confined to such a limited range of latitude. As yet, no points have been found near the sun's poles permanent and definite enough to permit precise observations covering a sufficient interval of time. Attempts have been made to use faculæ for the purpose, but they have proved too unstable and transitory.

By a discussion of all his observations, more than 5,000 in number, of 954 different groups of spots, Mr. Carrington deduced the expression $X = 865' - 165' \sin^{\frac{7}{4}} l$ for the daily motion of the surface of the sun in different solar latitudes, l representing the latitude in the formula, and X the daily motion in minutes of solar longitude. This, as was said before, would make the rotation period at the sun's equator a little less than 25 days. The expression, however, is purely empirical, and no imaginable theoretical explanation can be given for the fractional exponent $\frac{7}{4}$.

Faye, assuming on theoretical grounds that this exponent ought to be 2, finds from the same observations the formula $X = 862' - 186' \sin^2 l$, an expression which agrees with all but a few of the observations nearly as well as Carrington's.*

Spoerer, from observations of his own, made between 1862 and 1868, and combined with those of Secchi and others, derives the still different formula, $X = 1011'' - 203' \sin(41° 13' + l)$.

Finally, Zöllner, assuming, what no one else admits, that the surface of the sun consists of a thin liquid sheet circulating over a solid crust, deduces the formula

$$X = \frac{863' - 619' \sin^2 l}{\cos l}.$$

Either of these formulæ agrees very fairly with the facts observed; neither of them can be regarded as logically established upon a sound physical explanation.

The cause of this peculiar surface-drift is not yet known. Sir John Herschel was disposed to attribute it to the impact of meteoric matter striking the sun's surface mainly in the neighborhood of the equator, and so continually accelerating its rotation, as a boy's peg-top is whipped up by the skillfully applied lash. Perhaps there is nothing absurd in the idea that a sufficient

* Tisserand, from observations of 325 spots in 1874-'75, deduces the expression $X = 857' \cdot 6 - 157' \cdot 3 \sin^2 l$. This is probably less reliable than either of the preceding, being founded on a much smaller number of observations. We give it chiefly as an illustration of the amount of uncertainty still connected with the subject.

quantity of meteoric matter may reach the sun, or that the meteors move, for the most part, in the plane of the sun's equator, and direct, i. e., *with* and not *against* the motion of the planets—so that their fall would be mostly confined to the equatorial regions, and would thus hasten, and not retard, the surface motion. If this be so, the duration of the sun's rotation period should continually grow shorter, an effect which does not appear from a comparison of Scheiner's results with those most recently obtained. Of course, it may be that such an acceleration has actually occurred, only too small to be yet detected; still, it would seem probable that any "driving," sufficient to establish nearly two days' difference between the rotation periods at the equator and at latitude 40°, must have produced a very sensible effect within three hundred years.

It is more probable that the equatorial acceleration is connected in some way with the exchange of matter which, if the sun is for the most part gaseous, as now seems likely, must continually be going on between the outside and inside of the globe. If the photosphere is formed of masses *falling*, such an effect would be a necessary consequence. If we suppose that the outrushing streams of heated gas and vapor, as they rise, continue in the gaseous condition until they reach the summit of their ascent, and remain at this height long enough to acquire sensibly the rotation velocity corresponding to their altitude, and that then the products of condensation, resulting from their cooling, fall downward, and thus falling constitute the photosphere, we should have precisely the actual phenomenon. The rotation velocity of each visible element of the photosphere would be that corresponding to a greater altitude, and therefore greater than that naturally belonging to its

SUN-SPOTS AND THE SOLAR SURFACE.   137

observed position, and this difference would vary from the equator, where it would be a maximum, to the poles, where it would vanish.

Of course, it is not necessary to such an effect that the conditions supposed should be rigidly complied with; it will suffice to admit that in the photosphere the falling masses are more conspicuous than those which are ascending or stationary, and it would seem hardly possible that it should be otherwise. Whether, however, the effect thus produced would account in measure as well as kind for the observed phenomena, is a question requiring for its answer a more thorough mathematical investigation than the writer has yet been able to undertake.*

---

* A calculation, made since the text was written, leaves rather doubtful the correctness of the theory here suggested, *if we are to consider the motion of the spots as identical with that of the photosphere in which they seem to float*. According to Laplace's formula, slightly modified, and neglecting the resistance of the air, a body falling from a height, reaches the surface at a point east of the vertical by the quantity $\frac{2}{3} \frac{\pi}{\tau} g t^3$; in which expression π is 3·1416; τ is the length of the day in seconds; g, the measure of force of gravity, or twice the distance a body falls in one second; and t, the number of seconds occupied in falling. From this formula we find by differentiation that the eastward velocity of the body when it reaches the ground is $V = 2\frac{\pi}{\tau} g t^2$, or $\frac{4\pi}{\tau} h$ (since $h = \frac{1}{2} g t^2$), where h is the height fallen from. Taking twenty-seven days as the period of rotation at the sun's poles, we thus find that a fall of 5,000 miles would generate a relative eastward velocity of about 142 feet per second at the sun's equator, and this would give an apparent equatorial rotation of 25·8 days, the acceleration being only about two thirds enough to account for the observed facts, even if we neglect, as it would not be safe to do, the resistance of the solar atmosphere. If we consider *only the spots*, it would seem entirely possible that they may be produced by matter which has fallen from a height of even fifteen or twenty thousand miles, and that fall would be quite sufficient to account for their whole acceleration. It is an interesting question, therefore, whether the spots have or have

The fact that rapid changes in the configuration of a spot are generally accompanied by an eastward rush of the whole, favors the idea that a downfall of something from above is concerned in the matter.

The idea of Faye appears to have been nearly the reverse of that here suggested. He attributes the formation of the photosphere to gaseous matter not falling from above, but *ascending from below*, and starting from a stratum at a certain depth below the surface; by supposing the depth of this stratum to vary with the latitudes, being greatest at the poles of the sun and least at the equator, it is easy to explain on this hypothesis the accelerated motion of the surface at the equator, and to justify his formula, which makes the retardation at higher latitudes proportional to the square of the sine of the latitude; but no reason is evident why the depth of this stratum should vary.

As to Zöllner's idea that the equatorial acceleration is due to the friction between a liquid sheet, constituting the photosphere, and a solid nucleus below, it is hardly necessary to say that this view is in complete opposition to those held by almost all astronomers, and seems to be untenable in its fundamental assumptions.

The plane of the sun's rotation is slightly inclined to that of the earth's orbit. According to Carrington, the angle is 7° 15′, while Spoerer makes it 6° 57′. This plane cuts the ecliptic at two opposite points called the nodes, one of which is in longitude 73° 40′, according to Carrington, and 74° 36′, according to Spoerer. The axis of the sun is therefore directed to a point in the constellation of Draco, not marked by any conspicuous

---

not a forward motion with reference to the photospheric granules in their neighborhood? The writer knows of no existing observations or measurements which are capable of settling it.

star. Astronomers define its position by saying that its right ascension is $18^h\ 44^m$, and its declination is $64°$. It is almost exactly half-way between the bright star *a* Lyræ and the polar star.

The earth passes through the two nodes on or about the 3d of June and the 5th of December. At these times the spots move apparently in straight lines across the sun's disk, and his poles are situated on its circumference. During the summer and autumn, from June to December, the sun's northern pole is inclined toward the earth; during the winter months, the southern. The angle which the sun's axis appears to make with a north and south line in the sky (technically, the *position-angle* of the sun's axis) changes considerably during the year, varying $26°$ each side of zero. As it is often very desirable for an amateur to know this angle approximately, we insert the following little table, giving the position angle of the sun's *north* pole referred to the center of the disk. The table is derived from the much more extensive one in Secchi's "Le Soleil":

POSITION ANGLE OF SUN'S AXIS.

January 4, July 6................................$0°·00$.

Jan. 15, June 25....	5° west.	Dec. 24, July 17...	5° east.
Jan. 26, June 14....	10° west.	Dec. 15, July 29...	10° east.
Feb. 7, June 2......	15° west.	Dec. 3, Aug. 11....	15° east.
Feb. 22, May 18....	20° west.	Nov. 19, Aug. 27...	20° east.
March 18, April 25..	25° west.	Oct. 29, Sept. 20...	25° east.
April 5.............	26° 20' west.	Oct. 10...........	26° 20' east.

It is understood, of course, that the table is only approximate, because the numbers change slightly according to the place of the current year in the leap-year cycle; but the results obtained from it are always correct within about $\frac{1}{4}°$, which is near enough for most purposes.

After making due allowance for the equatorial acceleration, it is found that almost every spot has more or less motion of its own. Between latitudes 20° north and 20° south, Mr. Carrington finds, on the whole, a slight tendency to motion toward the equator, the movement amounting to a minute or two of arc *per diem;* from 20° to 30° on both sides of the equator, there is a somewhat more decided motion toward the poles. Faye has also shown that many spots move in small ellipses upon the surface of the sun, completing their circuits in a day or two, and repeating them with great regularity for weeks, and even months. Whenever a spot is passing through sudden changes, it generally moves forward upon the solar surface, as has already been mentioned, with something like a leap; and, when a spot divides into two or more, the parts generally separate with a very considerable velocity, as if (we do not say *because*) there was a repulsion between them.

The sun-spots, as has been already said, are not distributed over the sun's surface with anything like uniformity. They occur mainly in two zones on each side of the equator and between the latitudes of 10° and 30°. On the equator itself they are comparatively rare; there are still fewer beyond 35° of latitude, and only a single spot has ever been recorded more than 45° from the solar equator—one observed in 1846 by Dr. Peters (now of Hamilton College, then in Naples).

The figure shows the distribution of 1,386 spots observed by Carrington. The figure is constructed in this way: The circumference of the sun, on the left-hand side of the figure, is divided into five-degree spaces from the equator each way, and at each of them is erected a radial line whose length in *four-hundredths of an inch* is proportional to the number of spots ob-

SUN-SPOTS AND THE SOLAR SURFACE. 141

served within 2½° of latitude on each side. Thus, the line drawn at 20° north latitude, and marked "151," is $\frac{151}{400}$ of an inch long, and means that 151 spots were recorded between 17½° and 22½° north latitude.

It is at once evident from mere inspection that the distribution follows no simple law of latitude. On the northern hemisphere, the distribution, during the eight years over which the observations extend, was not very irregular, though there is a distinct minimum at 15°,

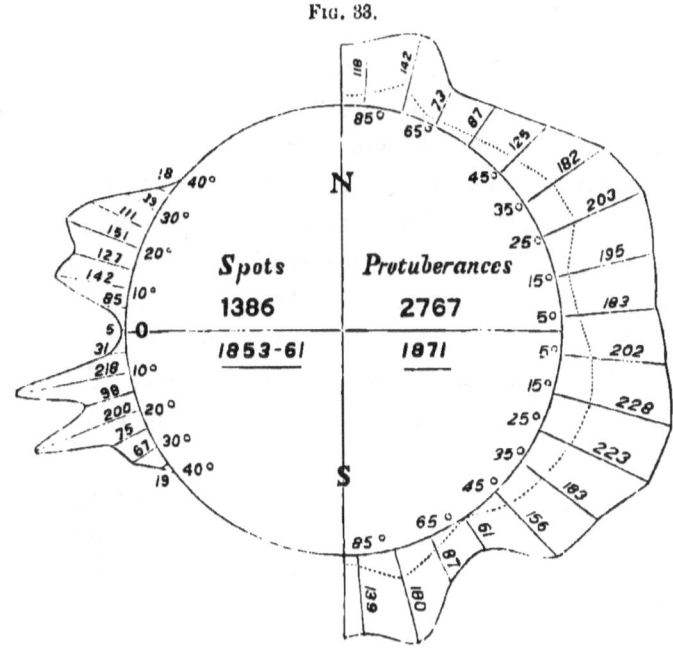

Fig. 33.

DISTRIBUTION OF SUN-SPOTS AND PROTUBERANCES.

and two maxima at about 11° and 22° of latitude. On the southern hemisphere the minimum at 15° is very marked, and the numbers at 10° and 20° are far in excess of those in the northern hemisphere. Of the whole number, 711 were in the southern hemisphere, as against 675 in the northern.

It is probable that this minimum at 15° of latitude and this difference between the two hemispheres are merely accidental and special to the eight years in question, as the observations of Spoerer from 1861 to 1867 show nothing of the kind.* It is to be noticed, moreover, that, at times when spots are abundant, their mean latitude is greater than when they are few, or, in other words, an increase in the number of spots generally carries with it a widening of the zones in which the spots appear. All the observations concur in showing this.

The cause of this distribution of the spots in zones is not known. It is probably connected with the origin of the spots themselves, and very possibly has something to do with the law of surface-motion just discussed. At least it is certain, as Faye pointed out some years ago, that, while at the solar poles and equator adjoining portions of the photosphere have no relative motion with reference to each other, yet in the middle latitudes this is not true; here each element of the surface has a different velocity from those immediately north and south of it, so that they drift by each other like the filaments of a liquid current which is suffering retardation, producing, as Faye supposes, whirlpools and eddies which, according to his view, generate the spots.

* Spoerer's observations, from 1861 to 1867, show the following distribution of 1,053 spots in latitude, viz.: + 35°, 4; + 30°, 4; + 25°, 16; + 20°, 50; + 15°, 133; + 10°, 198; + 5°, 114—in all 519 spots north of the solar equator. 40 spots were on the equator, or within 2° of it. South of the equator we have, in latitude: — 5°, 113; — 10°, 206; — 15°, 109; — 20°, 38; — 25°, 19; — 30°, 7; — 35°, 1; — 40°, 1—in all, 494 southern spots. In 1866, a year of spot minimum, there were only 94 spots in all, and of these 94, all but two were situated within 17° of the equator.

It is a question of much theoretical importance whether spots do or do not appear repeatedly at the same points; for, if this is really the case, it would make it almost certain that below the photosphere there must be a coherent nucleus, carrying with it in its rotation such volcanic or otherwise peculiar regions as to cause the breaking out of spots above them. There would be no difficulty in accounting for two or three dissolutions and reappearances in the same region without any such hypothesis, since a great disturbance in the solar atmosphere would not subside entirely for a long time. But, if it should turn out that, for many years, spots had, over and over again, broken out at the same point, the case would be changed. Spoerer seems rather disposed to hold that this is the fact, and there is a good deal in his observations, and also in those of others, to support his view; but, on the whole, considering the uncertainty of our knowledge of the true period of the sun's rotation, the evidence is not sufficient to establish it. If it should be shown to be true hereafter, it would compel an entire remodeling of the received views of the constitution of the sun.

CHAPTER V.

PERIODICITY OF SUN-SPOTS; THEIR EFFECTS UPON THE EARTH, AND THEORIES AS TO THEIR CAUSE AND NATURE.

Observations of Schwabe.—Wolf's Numbers.—Proposed Explanations of Periodicity.—Connection between Sun-Spots and Terrestrial Magnetism.—Remarkable Solar Disturbances and Magnetic Storms.—Effect of Sun-Spots on Temperature.—Sun-Spots, Cyclones, and Rainfall.—Researches of Symons and Meldrum.—Sun-Spots and Commercial Crises.—Galileo's Theory of Spots.—Herschel's Theory.—Secchi's First Theory.—Zöllner's.—Faye's.—Secchi's Later Opinions.—Other Theories.

It was early noticed that the number of sun-spots is very variable, but the discovery of a regular periodicity in their number dates from 1851, when Schwabe, of Dessau, first published the result of twenty-five years of observation. During this time he had examined the sun on every clear day, and had secured an almost perfect record of every spot that appeared upon the solar surface. He began his work without any idea of obtaining the result he arrived at, and says of himself, that, "like Saul, he went to seek his father's asses, and found a kingdom." His observations showed unmistakably that there is a pretty regular increase and decrease in the number of sun-spots, the interval from one maximum to the next being not far from ten years. Subsequent observations and a thorough examination of all known former records fully confirm this conclusion, except that the mean period appears to be somewhat greater, eleven and one ninth years being the value at present generally received. Professor R. Wolf, of Zürich, has been especially indefatigable in his investigations upon this subject, and has succeeded in disinterring from all sorts of hiding places a nearly complete history of the solar surface for the past hundred and fifty years. Among other things he finds among the unpublished manuscripts of Horrebow (a Danish astronomer who flourished a century ago) a distinct intimation (in 1776) that zealous and continued observation of the sun-spots might lead to "the discovery of a period, as in the motions of the other heavenly bodies," with the added remark that "then, and not till then, it will be time to inquire in what manner the bodies which are ruled and illuminated by the sun are influenced by the sun-spots"—alluding, perhaps, to certain ideas then, as now, more or less current, and illustrated by the attempt of Sir W. Herschel, a few years later, to establish a relation between the price of wheat and the number of sun-spots.

Wolf has brought together an enormous number of observations, and with immense labor has combined them into a consistent whole, deducing a series of "relative numbers," as he calls them, which represent the state of the sun as to spottedness for every year since 1745. His "relative number" is formed in rather an arbitrary manner from the observation of the spots: representing this number by r, the formula is, $r = k(f + 10g)$, in which g is the number of groups and isolated spots observed, and f the total number of spots which can be counted in these groups and singly, while k is a coefficient which depends upon the observer and his telescope. Wolf takes it as unity for

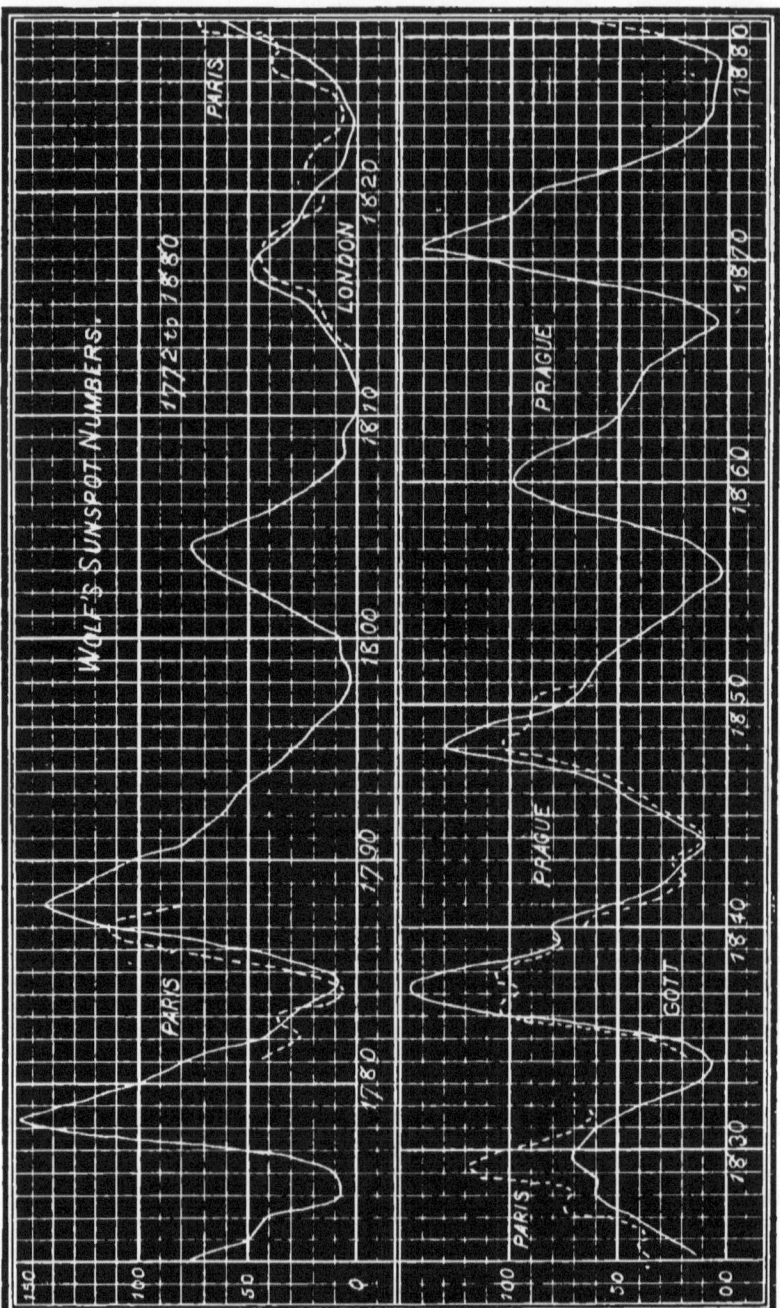

Fig. 34.—Wolf's Sun-Spot Numbers.

himself, observing with a three-inch telescope and power of 64. For an observer with a larger instrument, k would be a smaller quantity, while a less powerful instrument and less assiduous observer would receive a "k" greater than unity, as probably seeing fewer spots than Wolf himself would reach with his instrument. These relative numbers, as tested by the most recent photographic results of De La Rue and Stewart, are found to be quite approximately proportional to the area covered by the spots.

We give on the opposite page a figure deduced from the numbers, published by Wolf in 1877, in the "Memoirs of the Royal Astronomical Society," and showing their course year by year since 1772. The horizontal divisions denote years, and the height of the curve at each point gives the "relative number" for the date in question. For example, in 1870, about the middle of the year, the relative number was 140, while early in 1879 it ran as low as 3.

The dotted lines are curves of magnetic disturbance, with which at present we have no concern. Our diagram, on account of the smallness of the page, only goes back to 1772, but Wolf's investigations reach to 1610, and he gives, in the paper from which were derived the numbers used in constructing our diagram, the following important table of the maxima and minima of sun-spots since that date, dividing the results into two series, the first of which, from the paucity of observations, is to be considered of much inferior weight to the second:

148   THE SUN.

First Series.		Second Series.	
Minima.	Maxima.	Minima.	Maxima.
1610·8	1615·5	1745·0	1750·3
8·2	10.5	10·2	11·2
1619·0	1626·0	1755·2	1761·5
15·0	13·5	11·3	8·2
1634·0	1639·5	1766·5	1769·7
11·0	9·5	9·0	8·7
1645·0	1649·0	1775·5	1778·4
10·0	11·0	9·2	9·7
1655·0	1660·0	1784·7	1788·1
11·0	15·0	13·6	16·1
1666·0	1675·0	1798·3	1804·2
13·5	10·0	12·3	12·2
1679·5	1685·0	1810·6	1816·4
10·0	8·0	12·7	13·5
1689·5	1693·0	1823·3	1829·9
8·5	12·5	10·6	7·3
1698·0	1705·5	1833·9	1837·2
14·0	12·7	9·6	10·9
1712·0	1718·2	1843·5	1848·1
11·5	9·3	12·5	12·0
1723·5	1727·5	1856·0	1860·1
10·5	11·2	11·2	10·5
1734·0	1738·7	1867·2	1870·6
		(11·7)*	
		(1878·9)*	
Mean period.	Mean period.	Mean period.	Mean period.
11·20 ± 2·11	11·20 ± 2·06	11·16 ± 1·54	10·94 ± 2·52
± 0·64	± 0·63	± 0·47	± 0·76

From these data, Wolf derives a mean period of 11·111 years, with an average variability of 2·03 years, and an uncertainty of 0·307, due chiefly to the difficulty of fixing the precise date of maximum or minimum.

* The date 1878·9 and the corresponding period 11·7 are taken from a note by Wolf, in vol. xcvi of the "Astronomische Nachrichten." Wolf's mean numbers below (11·16, etc.) have not, however, been altered so as to take this last minimum into account, but stand as originally given in 1877.

A moment's inspection of the table shows that the period is not at all fixed and certain like that of an orbital motion, but is subject to great variations. Thus, between the maxima of 1829·9 and 1837·2 we have an interval of only 7·3 years, while between 1788 and 1804 it was 16·1 years.* A portion of this great variableness of period may, perhaps, be due to the incompleteness of our observations, but only a portion. It is quite likely that a fluctuation of much longer period, not far from fifty years, is, to some extent, responsible for the effect by its superposition upon the principal (eleven-year) oscillation.

Another important fact is that the interval from a minimum to the next following maximum is only about 4½ years on the average, while from the maximum to the next following minimum the interval is 6·6 years. The disturbance which produces the sun-spots springs up suddenly, but dies away gradually.

There is no question of solar physics more interesting or important than that which concerns the cause of this periodicity, but a satisfactory solution remains to be found. It has been supposed by astronomers of very great authority that the influence of the planets in some way produces it. Jupiter, Venus, and Mercury have been especially suspected of complicity in the matter, the first on account of his enormous mass, the others on account of their proximity. De La Rue and Stewart deduce from their photographic observations of sun-spots, between 1862 and 1866, a series of numbers, which strongly tend to prove that, when two of the powerful planets are nearly in line as seen from

---

* Some astronomers contend that there ought to be another maximum inserted about 1795. Observations about this time are few in number and not very satisfactory.

the sun, then the spotted area is much increased. They have investigated especially the combined effect of Mercury and Venus, Jupiter and Venus, and Jupiter and Mercury, as also the effect of Mercury's approach to, or recession from, the sun. In all four cases there seems to be a somewhat regular progression of numbers, though much less decided in the third and fourth than in the first and second. The irregular variations of the numbers are, however, so large, and the duration of the observations so short, that it is hardly safe to build heavily upon the observed coincidences, since they may be merely accidental. An attempt to connect the eleven-year period with that of the planet Jupiter also breaks down. While, for a certain portion of the time, there is a pretty good agreement between the sun-spot curve and that which represents the varying distance of Jupiter from the sun, there is complete discordance elsewhere. About 1870 the maximum spottedness occurred when the planet was *nearest* the sun, but at the beginning of the century the reverse was the case. Loomis (who is in favor of inserting a sun-spot maximum in 1794, and, on this hypothesis, deduces a mean sun-spot period of 10 years in place of 11·1) suggests that the conjunctions and oppositions of Jupiter and Saturn may be at the bottom of the matter. These occur at intervals of 9·93 years, from a conjunction to an opposition, or *vice versa*. But, when we come to test the matter, we find that, in some cases, sun-spot minima have coincided with this allineation of the two planets; in other cases, maxima.

It is, indeed, very difficult to conceive in what manner the planets, so small and so remote, can possibly produce such profound and extensive disturbances on the sun. It is hardly possible that their gravitation can be the agent, since the tide-raising power of Venus upon the solar surface would be only about $\frac{1}{750}$ of that which the sun exerts upon the earth; and in the case of Mercury and Jupiter the effect would be still less, or about $\frac{1}{1000}$ of the sun's influence on the earth. The sun (apart from the moon) raises a tide on the deep waters of the earth's equator, something less than a foot in elevation, so that, making all allowances for the rarity of the materials which compose the photosphere, it is quite evident that no planet-lifted tides can directly account for the phenomena. If the sun-spots are due in any way to planetary action, this action must be that of some different and far more subtile influence.

Several astronomers, among others Professor B. Peirce, seem to have adopted an idea before alluded to—first suggested, we believe, by Sir John Herschel—that the spots are caused by meteors falling upon the sun. According to this view, the periodicity of the spots could be simply accounted for by supposing the meteors to move in a very elongated orbit, with a period of 11·1 years, adding the additional hypothesis that at one part of the orbit they form a flock of great density, while elsewhere they are sparsely distributed. This meteoric orbit would have to lie nearly in the plane of the sun's equator, and have its aphelion near the orbit of Saturn. Of course there is no necessity to limit our hypothesis to a single meteor-stream. What we know of meteor-showers encountered by the earth, makes it likely that there may be several, of different periods; and thus we may account for some of the observed irregularities of the sun-spot period. The hypothesis has many excellent points, and we shall have occasion to recur to it again. At the same time, it may be said here that it seems very difficult to make it explain the enormous dimensions and persistence of many sun-spot groups, and the distribution of the spots on the sun's surface in two parallel zones, with a minimum at the equator. The irregularity in the epochs of maxima and minima is also much greater than would have been expected.

On the whole, it seems rather more probable that the periodicity is in the sun itself, depending upon no external causes, but upon the constitution of the photosphere and the rate at which the sun is losing heat. Perhaps we may compare small things with great by referring to the periodic explosions of the Icelandic geysers, or the "bumping" of ether and many other liquids in a chemist's test-tube. Looking at it in this light, we should imagine the course of events to consist of a gathering of deep-lying forces during a season of external quiescence, followed by an outburst, which relieves the internal fury; the rest and the paroxysms recurring, at somewhat regular intervals, simply because the forces, materials, and conditions involved, change only slowly with the lapse of time.

If such be really the case, it is clear, of course, that this periodicity is never likely to be very regular, and will not long keep step with any planetary march. Time of itself, therefore, will by-and-by solve the problem for us, or at least will refute any false hypothesis resting upon the recurrence of planetary positions.

Even more important than the problem of the cause of sun-spot periodicity, is the question whether this periodicity produces any notable effects upon the earth, and, if so, what? In regard to this question the astronomical world is divided into two almost hostile camps, so decided is the difference of opinion, and so sharp the discussion. One party holds that the state of the sun's surface is a determining factor in our terrestrial meteorology, making itself felt in our temperature, barometric pressure, rainfall, cyclones, crops, and even our financial condition, and that, therefore, the most careful watch should be kept upon the sun for economic as well as scientific reasons. The other party contends that there is, and can be, no sensible influence upon the earth produced by such slight variations in the solar light and heat, though, of course (excepting only the French astronomer Faye, so far as the writer knows), they all admit the connection between sun-spots and the condition of the earth's magnetic elements. It seems pretty clear that we are not in a position yet to decide the question either way; it will take a much longer period of observation, and observations conducted with special reference to the subject of inquiry, to settle it. At any rate, from the data now in our possession, men of great ability and laborious industry draw opposite conclusions

It certainly is not so plain that the sun-spots have not the influence which their worshipers, I had almost called them, claim for them, as to absolve us from the duty of investigating the matter in the most thorough manner. On the other hand, it is also by no means certain that we shall find the labor of investigation fruitful in precisely the manner and degree desired. Those who search for truth with honest endeavor may, nevertheless, be sure of their reward in some way.

I have said that there is no doubt as to the connection between the sun-spots and terrestrial magnetism.

In 1850, Lamont, of Munich, called attention to the

fact that the average daily excursions of the magnetic needle have a period which, from the few decades of observation at his command, he fixed at ten and one third years. Perhaps a word of explanation is needed here. Every one knows that the compass-needle does not point exactly north, and its divergence from the true meridian is different in different places. On the Atlantic coast of the United States, for instance, the north pole of the magnet points west of north, and on the Pacific coast east of north. What is more: at any particular place the direction of the needle is continually changing, these changes being like the changes in the temperature of the air, in part regular and predictable, and partly lawless, so far as we can see. One of the most noticeable of the regular magnetic changes is the so-called diurnal oscillation; during the early part of the day, between sunrise and one or two o'clock P. M., the north pole of the needle moves toward the west in these latitudes, returning to its mean position about 10 P. M., and remaining nearly stationary during the night. The extent of this oscillation in the United States is about 15' of arc in summer, and not quite half as much in winter; but it differs very much in different localities and at different times, and also—and this is Lamont's discovery—the average extent of this diurnal oscillation at any given observatory increases and decreases pretty regularly during a period of $10\frac{1}{3}$ years, according to his calculations. As soon as Schwabe announced his discovery of the periodicity of the solar spots, Sabine in England, Gautier in France, and Wolf in Switzerland, at once and independently perceived the coincidence between the spot-maxima and those of the magnetic oscillation. Faye has recently attempted to impugn this conclusion. In order to make his point, he insists that the magnetic maximum is shown by Cassini's observations to have occurred early in 1787, and, dividing the interval between this and the last magnetic maximum, near the close of 1870, by 8, the number of intervening periods, he gets 10·45 years for the mean magnetic period, instead of 11·11. The reply is, that the observations both of the sun-spots and of the magnetic elements near the close of the eighteenth century are so meager and unsatisfactory that the evidence as to the precise time of maxima and minima is very incomplete. It is even doubtful, as has been said before, whether there should not be recognized an additional sun-spot maximum in 1795, over and above those enumerated by Wolf.

The convincing evidence as to the reality of the asserted connection lies in the closeness with which, ever since we have been in possession of continuous and satisfactory observations, the magnetic curve copies that of the sun-spots. In Fig. 34 the dotted curves represent the mean amount of magnetic oscillation as deduced by Wolf from various series of observations. Since 1820 the record is almost continuous, and the coincidence of the curve is such as to leave no doubt in an unprejudiced mind.*

The argument is much strengthened by an examination of records of the aurora borealis. Occasionally so-called "magnetic storms" occur, during which the compass-needle is sometimes almost wild with excitement, oscillating 5° or even 10° within an hour or two. These "storms" are generally accompanied by an aurora, and

---

* A discussion, by Balfour Stewart, of the observations at Kew, between 1856 and 1867, brings out the correspondence very beautifully, and seems to show that the magnetic changes lag behind the sun-spots about five months.

an aurora is *always* accompanied by magnetic disturbance.

Now, when we come to collate aurora observations with those of sun-spots, as Loomis has done with great care and thoroughness, we find an almost perfect parallelism between the curves of auroral and sun-spot frequency.

It is not easy to frame any satisfactory theory to account for this effect of solar disturbances upon our terrestrial magnetism. The connection can hardly be in the way of temperature, for the influence of sun-spots in this respect is so slight that it is still an open question whether we do or do not get from the sun more than the average amount of heat during a sun-spot maximum. Probably the magnetic connection is more immediate and direct; perhaps in some way kindred with the action which drives off the material of a comet's tail, and proves that other forces besides gravitation are operative in inter-planetary space.

There are a number of observed instances which, though not sufficient to demonstrate the fact, still render it very probable that every intense disturbance of the solar surface is propagated to our terrestrial magnetism with the speed of light. The occurrence observed by Carrington and Hodgson (p. 119), on September 1, 1859, was immediately followed by a magnetic storm of unusual intensity, the auroral displays being most magnificent on both sides of the Atlantic, and even in Australia. Another instance fell under the writer's notice in the course of a series of spectroscopic observations at Sherman. On August 3, 1872, the chromosphere in the neighborhood of a sun-spot, which was just coming into view around the edge of the sun, was greatly disturbed on several occasions during the forenoon. Jets of luminous matter of intense brilliance were projected, and the dark lines of the spectrum were reversed by hundreds for a few minutes at a time. There were three especially notable paroxysms at 8.45, 10.30, and 11.50 A. M. local time. At dinner the photographer of the party, who was making our magnetic observations, told me, before knowing anything about what I had been observing, that he had been obliged to give up work, his magnet having swung clear off the scale. Two days later the spot had come around the edge of the limb. On the morning of August 5th I began observations at 6.40, and for about an hour wit-

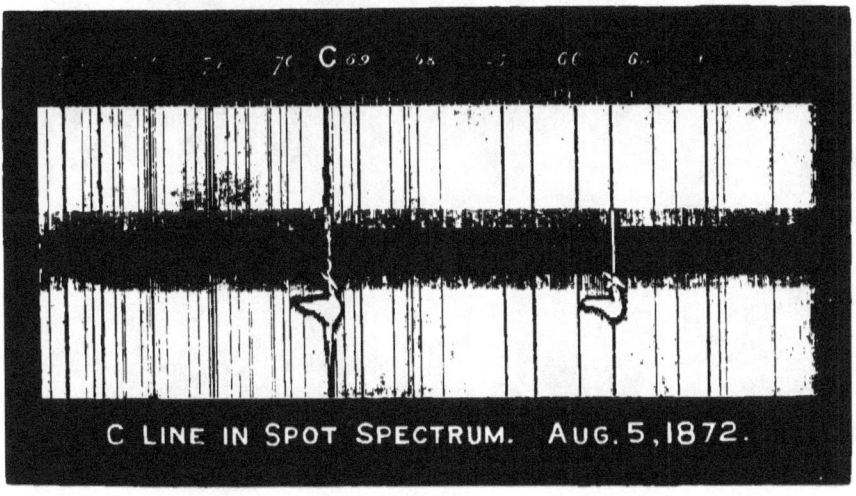

Fig. 35.

C LINE IN SPOT SPECTRUM. AUG. 5, 1872.

nessed some of the most remarkable phenomena I have ever seen. The hydrogen-lines, with many others, were brilliantly reversed in the spectrum of the nucleus, and at one point in the penumbra the C line sent out what looked like a blowpipe-jet, projecting toward the upper end of the spectrum, and indicating a motion along the line of sight of about one hundred and twenty miles per second. This motion would die out and be renewed again at intervals of a minute or two. The figure gives an idea of the appearance of the spectrum. The disturbance ceased before eight o'clock, and was not renewed that forenoon. On writing to England, I received from Greenwich and Stonyhurst, through the kindness of Sir G. B. Airy and Rev. S. J. Perry, copies

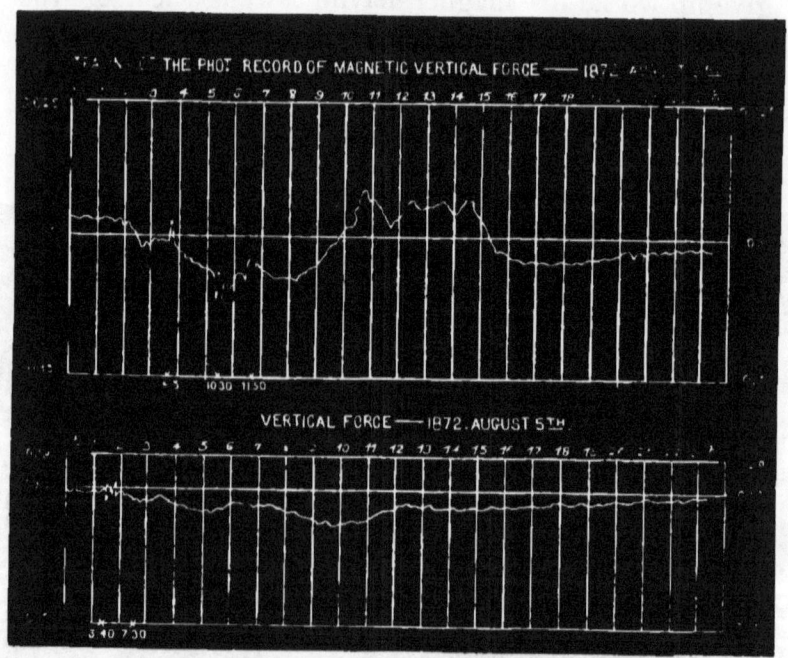

Magnetic Curves at Greenwich (August 3 and 5, 1872).

of the photographic magnetic records for those two days. Fig. 36 is reduced from the Greenwich curve. That obtained at Stonyhurst is essentially the same. It will be seen that on August 3d, which was a day of general magnetic disturbance, the three paroxysms I noticed at Sherman were accompanied by peculiar twitches of the magnets in England. Again, August 5th was a quiet day, magnetically speaking, but just during that hour when the sun-spot was active, the magnet shivered and trembled. So far as appears, too, the magnetic action of the sun was instantaneous. After making allowance for longitude, the magnetic disturbance in England was strictly simultaneous, so far as can be judged, with the spectroscopic disturbance seen on the Rocky Mountains.

Of course, as has been said, no two or three coincidences such as have been adduced are sufficient to establish the doctrine of the sun's immediate magnetic action upon the earth, but they make it so far probable as to warrant a careful investigation of the matter—an investigation, however, which is not easy, since it implies a practically continuous watch of the solar surface.

As to the effect of sun-spots upon terrestrial temperature, no conclusion seems possible at present. The spots themselves, as Henry, Secchi, Langley, and others have shown, certainly radiate to us less heat than the general surface of the sun. According to the elaborate determinations of Langley, the umbra of a spot emits about fifty-four per cent. and the penumbra about eighty per cent. as much heat as a corresponding area of the photosphere. The *direct* effect of sun-spots is, therefore, to make the earth cooler. As the total area covered by spots, even at the time of maximum, never exceeds $\frac{1}{500}$ of the whole surface of the sun,* it follows that *directly* they may diminish our heat-supply by about $\frac{1}{1000}$ of the whole. Whether this effect would be sensible or not, is a question not easily answered.

* There are a few cases on record where the area of a group of spots has much exceeded this figure for a few days, but the results of Stewart and De La Rue show that it is an outside estimate for the average spotted area during any year of sun-spot maximum.

But, while the direct effect would be of this nature, it is quite probable that it is at least fully compensated by another of the opposite character. We get our light and heat from the photosphere which is covered by an atmosphere of gases, and in this atmosphere a considerable absorption occurs. Now, if the level of the photospheric surface be disturbed, so that it is covered with waves and elevations of any considerable height, as compared with the thickness of the overlying atmosphere, then, as Langley has shown, the radiation will at once be increased; since, while the absorption is increased by a certain percentage for those portions of the photosphere which are depressed below their ordinary level, it is *much more decreased* for those that are raised.

The reason of this is that, when a luminous object is immersed in an absorbing medium, it loses much more light for the first foot of submergence than for the second, and more for the second than for the third; so that when it has reached a considerable depth it requires an additional submergence of many feet to diminish its radiation as much as the first foot did. If, therefore, sun-spots are accompanied by considerable vertical disturbance of the photosphere, as is almost certain, we must have as a result an increased radiation on account of the disturbance, offsetting, more or less entirely, the opposite effect which is at first view most obvious.

Then, again, it is altogether probable that spots are either due to, or accompanied by, an eruptive action— the internal, and hotter, gases bursting through the photosphere with unusual abundance during seasons of spot-maximum. This must necessarily tend to increase the emission of heat from the sun, and possibly by a considerable amount. But, on the other hand, any considerable increase in the thickness of the chromosphere, such as might result from abundant and long-continued eruption, would work in the opposite direction.

It is impossible, therefore, to predict, *a priori*, which effect will predominate, or to say whether the mean temperature of the earth ought to be raised or lowered during a sun-spot maximum; and thus far no comparison of observations has settled the matter to general satisfaction. At least, no longer ago than 1878, Balfour Stewart, who ought to know if any one, writes, " It is nearly, if not absolutely, impossible, from the observations already made, to tell whether the sun be hotter or colder, as a whole, when there are most spots on his surface."

On the one hand, Jelinck, from all temperature observations available in Germany up to 1870, found the influence of sun-spots entirely inappreciable, though from the same observations he did deduce minute effects produced by the changes in the distance and phase of the moon. On the other hand, Mr. Stone, while astronomer royal at the Cape of Good Hope, and Dr. Gould, in South America, consider that the observations taken at their stations show a distinct though slight *diminution of temperature* at the time of a sun-spot maximum: according to Dr. Gould the difference at Buenos Ayres between maximum and minimum amounts to about $1\frac{3}{4}°$ Fahr. At the Cape of Good Hope, Mr. Stone finds the difference to be about three fourths of a degree from thirty years' observations—at least, if we rightly interpret his curve of temperatures, for it is not quite clear what unit of temperature is used in constructing his diagram.

At Edinburgh, Piazzi Smyth finds in the records of the rock thermometers a marked eleven-year periodicity, of which the range amounts to about a degree (Fahr.), and the maxima, instead of coinciding with the sun-spot minima, come about two years behind them.

As against all these, Mr. F. Chambers, of Bombay, draws from the Asiatic observations of the *barometer*, between 1848 and 1876, the conclusion that the sun is hottest when most spotted. His paper will be found in " Nature " of September 26, 1878, with a diagram of the barometric curves from which he draws his conclusions.

On the whole, perhaps, as things now stand, it would be fair to say that there is a small balance of probability in favor of the statement that years of sun-spot maximum are a degree or so cooler than those of spot-minimum; but the balance is very slight indeed, and the next investigation of somebody else may carry it to the other side.

As regards the influence of sun-spots upon storms and rainfall, the evidence, if not entirely conclusive, as it is considered by Mr. Lockyer and some other high authorities, is at least considerably stronger. In 1872 Mr. Meldrum, director of the observatory at the Mauritius, published a comparison between the number of cyclones observed in the Indian Ocean and the state of the sun, and pointed out that the number of cyclones was greatest at the time of a sun-spot maximum. We quote his words (" Nature," vol. vi, p. 358): " Taking the maxima and minima epochs of the sun-spot period, and one year on each side of them, and comparing the number of cyclones in these three-year periods, we get the following results:

Years.		No. of cyclones in each year.	Total No. of cyclones.
Maxima	{ 1847... { 1848... { 1849...	4 6 5	15
Minima	{ 1855... { 1856... { 1857...	4 1 3	8
Maxima	{ 1859... { 1860... { 1861...	5 8 8	21
Minima	{ 1866... { 1867... { 1868...	5 2 2	9
Maxima	{ 1870... { 1871... { 1872...	3 4 7	14 "

Subsequently Mr. Meldrum made more extensive comparisons, including not only cyclones proper, but other great storms, and brings out essentially the same results. At the same time it is to be noted that the yearly numbers vary enormously, and, on referring to his second paper ("Nature," vol. viii, p. 495), it will be found that the number for the sun-spot maximum, 1847–'49, is only twenty-three, while that for the minimum, 1866–'68, is twenty-one. (Mr. Meldrum coaxes the first sun-spot maximum a little by using the years 1848–'50 in his comparison; rather unwarrantably, it would seem, since the epoch of spot-maximum was 1848·1: by using those years, he gets twenty-six instead of twenty-three.)

The variations from year to year are so extreme that it is sufficient to say that the observations can hardly be considered as demonstrative without much further confirmation from other sources.

Mr. Meldrum has attempted to supply this confirmation by tabulating the rainfall at a number of stations in and near the Indian Ocean. He obtains a result confirmatory on the whole, though there are several discrepancies. Mr. Lockyer, from observations of the rainfall at the Cape of Good Hope and Madras, gets corroborative figures. Mr. Symons, from the British rainfall of the past one hundred and forty years, gets an equivocal result. American stations, so far as they have been tested, are on the whole rather in opposition to those of the Indian Ocean, indicating somewhat less rain than usual during a sun-spot maximum. But, as any one can see by consulting Mr. Symons's paper in " Nature," vol. vii, pp. 143–145, in which he has tabulated an immense number of rainfall statistics, the evidence is extremely conflicting—altogether different in force and character from that which demonstrates the magnetic influence of solar disturbances.*

---

* Since writing the above, we have received from Mr. Meldrum his paper published in the " Monthly Notices of the Mauritius Meteorological Society," for December, 1878. In it he discusses at length the rainfall of more than fifty different stations in all parts of the earth, and also the levels of many of the principal European rivers. The discussion covers nearly all the available data from 1824 to 1867. It is only just to Mr. Meldrum to say that the treatment seems to be sufficiently thorough, perfectly fair, and the result of the whole is decidedly in favor of his opinion that there is a real connection between the annual rainfall and the state of the solar surface. He finds the average rainfall for the earth to be about 38·5 inches annually; the range between the maximum and minimum is about four inches; and the rainfall maximum occurs about a year after the sun-spot maximum, though with a good deal of variation at different stations. In some countries, indeed, and at some times (in the United States, for instance, between 1834 and 1843), the results conflict with the theory, but the general accordance is striking, and seems to warrant his concluding statement that " the mean rainfalls of Great Britain, the Continent of Europe, America, and India, as represented by all the returns that have been received, have, notwithstanding anomalies, varied directly as Wolf's sun-spot numbers have varied, and the epochs of maximum and minimum rain have nearly coincided with

Still other attempts have been made to establish a connection between sun-spots and various terrestrial phenomena. Thus, Dr. T. Moffat, in 1874, published results tending to show that in sun-spot years the average quantity of atmospheric *ozone* is somewhat greater than during a spot-minimum.

Another eminent physician, whose name escapes us, endeavored, a few years ago, to show that the visitations of Asiatic cholera are periodical, and that their period depends upon that of the sun-spots, being just *once and a half* as *long*—about fifteen years. This periodicity may be real, *perhaps;* but, if so, the fact that the cholera maxima are alternately synchronous with the maxima and minima of the spots, would be sufficient to disprove the idea of any causal connection between the phenomena.

The latest, and one of the most interesting, of the essays in this general direction, is that of Professor Jevons, who seeks to show a relation between sun-spots and commercial crises. The idea is by no means absurd, as some have declared—it is a mere question of fact. If sun-spots have really any sensible effect upon terrestrial meteorology, upon temperature, storms, and rainfall, they must thus indirectly affect the crops, and so disturb financial relations; in such a delicate organization as that of the world's commerce, it needs but a feather-weight, rightly applied, to alter the course of trade and credit, and produce a "boom" (if we may be forgiven the use of so convenient a word), or a crash.

We have not time or space to discuss Mr. Jevons's paper, but must content ourselves with saying that, to

those of the sun-spots. The rainfalls at five stations in the southern hemisphere, for shorter periods, give similar results."

us at least, the facts do not seem to warrant his conclusion. Mr. Proctor, in an article published in "Scribner's Magazine," for June, 1880, has gone over the subject very thoroughly and fairly.

It can do no harm to reiterate and emphasize what was said a few pages back, that the question of sun-spot influence can not be considered settled; and that the only method of deciding it is by a continuous series of careful observations, conducted specially for the purpose, or at least conducted with reference to the conditions of the problem, since the same observations would also be useful as data for various other investigations. We need, and ought to have, a continuous record of the state of the solar surface, such as it is hoped may be secured by the coöperation of the new astrophysical observatories at Potsdam and Meudon, with a few other sister institutions soon, we hope, to be organized in various parts of the world. To go with these solar observations we need also a system of simultaneous meteorological observations, representing both northern and southern, eastern and western hemispheres, so that the local may disappear from our mean results, leaving only the general and cosmical. Such a system we may reasonably hope to see established in the near future.

THEORIES AS TO THE CAUSE AND NATURE OF SUN-SPOTS.

Naturally, the remarkable phenomena of the sunspots have invited speculation as to their cause.

As has been mentioned already, some of the early observers believed the spots to be planetary bodies circulating around the sun, very near its surface. This opinion Galileo unanswerably refuted by pointing out that in that case the spot, in its movement around the sun, ought to be visible less than half the time. He, on the other hand, proposed the theory that they are clouds floating in the solar atmosphere.

This view, in one form or another, has since been held by many astronomers of great authority. Derham believed these clouds to be eruptions from solar volcanoes, and in our own times Capocci has adopted and maintained the same theory. Peters seems to have considered it favorably in 1846, at least so far as the volcanic part of the hypothesis is concerned, while Kirchhoff seems to have assented to Galileo's original opinion unmodified. If the statement be interpreted to mean that sun-spots are masses of cloudy matter, less luminous than the photosphere, and floating *in*, not *above*, the photosphere, probably a very large proportion of the students of solar physics would to-day agree to it. Galileo, however, believed the spot-clouds to be high above the shining surface, which we now know not to be the fact; for the observations of Wilson, in 1769, mentioned a few pages back, and the whole body of observations since then, have placed it beyond dispute that the umbra of a sun-spot lies several hundred miles below the level of the photosphere.

Lalande, however, was not disposed to accept Wilson's doctrine, and maintained that the sun-spots are the tops of solar mountains projecting above the luminous surface—islands in the ocean of fire. In this hypothesis the penumbra is accounted for by the shelving sides of the mountains seen through the semi-transparent flame. Of course, the observed motions of the spots, as well as the discovery of Wilson, are entirely inconsistent with this idea. It will be noticed that the theories already mentioned, as well as that of Sir William Herschel, which we must now present, all proceed upon the assumption that the central core of the sun is solid.

About the beginning of the present century, Sir William Herschel, after a careful study of the facts, but much influenced by the belief that the sun must (for theological reasons) be a habitable body, proposed an hypothesis which stood unchallenged for nearly half a century, and still maintains its place in some of our text-books of astronomy.

He supposed the central portion of the sun to be solid; its surface cool, non-luminous, and habitable. Around this he placed two envelopes of cloud—the outer one, the photosphere, incandescent, blazing with unimaginable fury; the inner one non-luminous, dark itself, but capable of reflecting light from its upper surface, and acting as a screen to protect the underlying country from the heat of the photosphere. The spots he supposed to be caused by holes temporarily opening

FIG. 37.

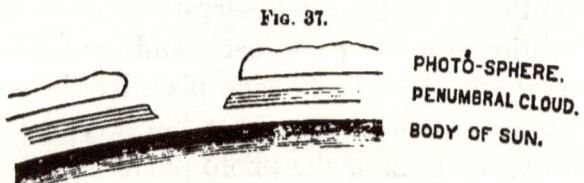

HERSCHEL'S SUN-SPOT THEORY.

in the clouds, through which we could look down upon the dark surface of the central globe; the penumbra being caused by the intermediate cloud-layer, opening less widely than the photosphere. The figure illustrates this theory. As to the cause of the openings he uttered no decided opinion, though suggesting that they might be due to volcanic eruptions, forcing their way up through the higher atmosphere.

His son, Sir John Herschel, many years later, proposed an explanation which would make the spots to be great whirling storms *boring down* through the photosphere and clouds, instead of eruptions pushing their way outward. According to him, the rotation of the sun causes an accumulation of the solar atmosphere at the sun's equator—a thickening of the layer which obstructs the radiation of heat. This being so, there should be on the sun, as on the earth, though for an entirely different reason, a temperature higher in the equatorial regions than elsewhere; and then would follow a long train of consequences, among them these: the solar atmosphere would be disturbed by currents like the tradewinds on the earth; there would be stormy zones on each side the equator, and these storms would furnish an explanation of the spots.

To a certain extent, the cause adduced must actually exist. The sun's rotation must necessarily thicken the atmospheric layer which overlies the photosphere (i. e., it must, if the surfaces of the photosphere and chromosphere can be regarded as *level* surfaces), and this cause must tend to raise the actual temperature of the sun's equator, while at the same time it must diminish its radiation to the earth, and so render the solar equator *apparently cooler*, as tested by our observations from the earth. But, so far as can be judged, this effect is quite insensible, as it should be, since the sun's rotation is so slow; and the motions of the spots show no such systematic drift north or south as solar trade-winds would necessarily produce.

The elder Herschel's theory satisfies all the telescopic appearances of sun-spots quite as well, perhaps, as any yet proposed. It breaks down in its assumption that the principal portion of the sun is a solid mass, an assumption which is now almost universally regarded as incompatible with what we know of the solar temperature, radiation, and constitution.

It seems to modern physicists an unavoidable conclusion that the sun's central mass must be gaseous, or at least not solid. Setting out with this idea, Faye and Secchi independently, about 1868, proposed the theory that the spots are openings in the photosphere, through

FIG. 38.

SECCHI'S FIRST-SPOT THEORY.

which the internal gases are bursting outward. According to this view, the umbra is dark, because the gaseous center of the sun, which is seen through the opening, has a lower radiating power than the incandescent droplets which compose the clouds of the photosphere. We present one of Secchi's figures illustrating this view. The theory is so simple that it is a pity it is not true.

But it was abandoned by its proposers as soon as it was clearly pointed out that in that case the spectrum of the umbra of a sun-spot should be composed of bright lines; and Secchi himself and others had shown that it is not so at all, but a spectrum due to increased absorption, and probably indicating, not an up-rush of heated gases through the photosphere, but a descent of cooler and less luminous matter from above. About 1870 Zöllner proposed a peculiar theory which has many good points about it, but seems obnoxious to fatal objections, and has found very few defenders. He conceives the surface of the sun to be *liquid*—a molten mass overlaid by an atmosphere of vapor. This liquid surface he imagines to be here and there covered at times by slag-like masses of much lower radiating power, the result of local cooling. Around their edges the solar flames burst out with redoubled fury, but at the center the cooler mass of scoria determines a downward current, so as to establish a powerful circulation in the solar atmosphere—downward at the center of the spot, outward in all directions at the surface of the slag, upward all around its margin, and inward, toward the center, in the upper air. This theory admirably agrees with the spectroscopic phenomena; but the hypothesis of a continuous liquid shell, cool enough to permit the formation of scoriæ, seems inconsistent with other phenomena, which make it impossible to admit so low a temperature at so great a depth.

At present, opinion, for the most part, seems to be divided between two rival theories proposed by Faye and Secchi.

Faye conceives the sun-spots to be the effect of solar storms; Secchi believes them to be dense clouds of eruption-products settling down into the photosphere *near*, but not *at*, the points where they were ejected.

Faye, it will be remembered, supposes the sun's peculiar law of rotation to be due to the hypothetical fact that the ascending masses of vapors (which form the photosphere by their condensation) start from a stratum whose depth below the visible surface regularly diminishes from the equator toward the poles. Hence result currents parallel to the equator, and the consequence is that, generally speaking, neighboring portions of the photosphere have a relative drift. At the equator and at the poles this drift vanishes, but is most considerable in the middle latitudes. Now, it is Faye's theory that, in consequence of this relative drift, eddies are formed, as explained on a preceding page; these eddies become cyclones or whirls precisely analogous to those seen in water where a rapid current is obstructed by an obstacle. In such a case, as every one knows, tunnel-shaped vortices are formed, down which floating materials and air are carried to considerable depths. Our terrestrial whirlwinds and tornadoes are produced, according to Faye (but in opposition to the generally received theories), in a similar manner, beginning from *above*, and penetrating downward until the point of the whirling vortex reaches and sweeps the earth. Now, such a vortex, on the solar scale, is the essence of a sunspot, according to Faye.

It is evident at once that this theory gives a reasonable explanation of the distribution of the spots in two parallel zones on each side of the sun's equator, and that the drifting action, in which the cause of the spots is supposed to lie, is a *vera causa*.

The theory accords very well, also, with the phenomena which accompany the subdivision of spots, since whirls in water and cyclones in the terrestrial atmosphere behave in precisely the same sort of way. It fairly meets, too, the spectroscopic indications. The cavity filled with descending vapors would naturally give just such a kind of spectrum as that which is ordinarily observed. Moreover, the gases carried down in the vortex below the photosphere, especially the hydrogen, would boil up again all around the whirlpool, and thus we could account for the ring of faculæ and prominences which, as a general rule, environs every spot of considerable magnitude. Some of the more obvious objections can also be easily disposed of. Thus, it has been said that, if the sun-spots are such vortices, they ought to be circular in outline. Faye replies that we see, not the vortex itself, but a great cloud of cooler gases, sucked down from above and gathered into the storm from all sides, and the form of this cloud would depend upon a multitude of circumstances.

But there are other objections which are not so easily met. It the theory be true, all spots are whirls and ought to show a vortical motion, and, what is more, all spots north of the equator ought to whirl in the same direction, and *against* the hands of a watch (as seen from the earth), while those in the sun's southern hemisphere should revolve in the contrary direction, precisely as cyclones do in the atmosphere of the earth.

Now, this is not the case at all. As we have seen, only a very small percentage of the spots show any trace of vorticose motion; and, so far from observing any uniformity in the direction of rotation on each side of the equator, we frequently find different members of the same group of spots, or even different portions of the self-same spot, revolving oppositely.

In fact, when we come to look into the matter numerically, we find that the *drift*, which Faye makes the determining factor of sun-spot genesis, is far too slight to produce such effects.

It is very easy to compute this drift if we assume the correctness of Faye's own formula for the motion of a point on the sun's surface in any given solar latitude, viz., $V' = 862' - 186' \sin^2 \lambda$; V' in this formula being the number of minutes of solar longitude passed over by any given point in twenty-four hours.

If we apply this formula to two points on the solar surface, one in latitude 20° and the other in latitude 20°·1', i. e., about 123 miles north of the first, we shall find that the first has a daily motion of 840·242' and the second 840·207', a difference of only ·035', or (in this latitude) 4·17 miles. That is to say, if we take two points on the solar surface, on the same meridian, in latitude 20°, at a distance of 123 miles, the one nearer the equator will, at the end of twenty-four hours, have drifted about $4\frac{1}{8}$ miles to the eastward of the other.

If we make the same calculation for latitude 45°, we get a result a trifle greater—about $4\frac{1}{3}$ miles per day.

With these figures it is easy to see why the sun-spots do not behave more like the disturbances of our terrestrial atmosphere, in exhibiting cyclonic motion as a regular and invariable characteristic, instead of an occasional and rather a rare phenomenon.

Secchi's latest theory is based essentially upon the idea, certainly borne out by observation, that eruptions are continually breaking through the photosphere, and carrying up metallic vapors from the regions beneath. He imagines that these vapors, after becoming considerably cooled, descend upon the photosphere and form depressions in it, which are filled with these less luminous and absorbent materials. It is difficult to see why the effect should remain so persistent, or why, even if the eruption be long maintained, the cloud should continue to descend in the same place. In fact, as was said only a few moments ago, a spot is generally surrounded by a ring of eruptions, and things take place as if they were all pouring their ejections into the same receptacle—as if there were, in fact, some such downward suction through the center of the spot as the theory of Faye supposes, an aspiration capable of drawing in toward the spot all erupted materials in the vicinity.

The writer some time ago suggested a modification of the theory, which may perhaps partly explain the facts. It may be that the spots are depressions in the photospheric level, caused not directly by the pressure of the erupted materials from above, but by the *diminution of upward pressure* from below, in consequence of eruptions in the neighborhood; the spots thus being, so to speak, *sinks* in the photosphere. Undoubtedly the photosphere is not a strictly continuous shell or crust, but it is *heavy* as compared with the uncondensed vapors in which it lies, just as a rain-cloud in our terrestrial atmosphere is heavier than the air, and it is probably continuous enough to have its upper level affected by any diminution of pressure below. The gaseous mass below the photosphere supports its weight and the weight of the products of condensation, which must always be descending in an inconceivable rain and snow of molten and crystallized material.. To all intents and purposes, though nothing but a layer of clouds, the photosphere thus forms a constricting shell, and the gases beneath are imprisoned and compressed. Moreover, at a high temperature the viscosity of gases is vastly increased, so that quite probably the matter of the solar nucleus resembles pitch or tar in its consistency more than what we usually think of as a gas. Consequently, any sudden diminution of pressure would propagate itself slowly from the point where it occurred. Putting these things together, it would seem that, whenever a free outlet is obtained through the photosphere at any point, thus decreasing the inward pressure, the result would be the sinking of a portion of the photosphere somewhere in the immediate neighborhood, to restore the equilibrium; and, if the eruption were kept up for any length of time, the depression in the photosphere would continue till the eruption ceased. This depression, filled with the overlying gases, would constitute a spot. Moreover, the line of fracture, if we may call it so, at the edges of the sink would be a region of weakness in the photosphere, so that we should expect a series of eruptions all around the spot. For a time the disturbance, therefore, would grow, and the spot would enlarge and deepen, until, in spite of the viscosity of the internal gases, the equilibrium of pressure was gradually restored beneath. So far as we know the spectroscopic and visual phenomena, none of them contradict this hypothesis. There is nothing in it, however, to account for the distribution of the spots in solar latitudes, nor for their periodicity. Perhaps the longitudinal drift, slight as it is, which Faye makes the foundation of his theory, may have some power to determine the region of eruptions. Possibly, too, there may be something in the belief that the fall of meteors produces the spots, an idea already referred to in connection with their periodicity. While it is hardly possible that, *directly*, a meteor, such as we know meteors upon the earth, could by its fall produce even a small sun-spot, it is not easy to say what might be the *indirect* effects consequent upon its passage through the photosphere, and its disturbance of the dynamical equilibrium.

Certainly, no theory of sun-spots can be considered complete which does not account for their distribution and periodicity, as well as the more obvious telescopic and spectroscopic phenomena; and it must be admitted that no theory yet proposed satisfactorily covers the whole ground.

Whatever may be their cause, however, it is probable that the annexed figure gives a fair idea of the arrangement and relations of the photospheric clouds

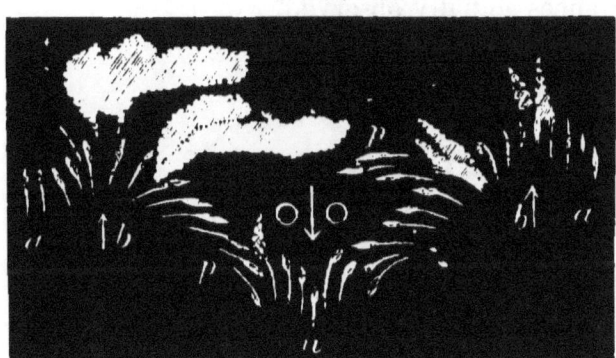

Fig. 30.

CONSTITUTION OF A SUN-SPOT.

in the neighborhood of a spot. Over the sun's surface generally, these clouds probably have the form of vertical columns, as at *a a*. Just outside the spot, the level of the photosphere is usually raised into faculæ, as at *b b*. These faculæ are for the most part overtopped by eruptions of hydrogen and metallic vapors, as indicated by the shaded clouds. Of these metallic eruptions we shall have more to say in the chapter upon the chromosphere and prominences, only remarking here that, while the great clouds of hydrogen are found everywhere upon the sun, these spiky, vivid outbursts of metallic vapors seldom occur, except just in the neighborhood of a spot, and then only during its season of rapid change. In the penumbra of the spot the photospheric filaments become more or less nearly horizontal, as at $p\,p$; in the umbra, at u, it is quite uncertain what the true state of affairs may be. We have conjecturally represented the filaments there as vertical also, but depressed and carried down by a descending current. Of course, the cavity $o\,o$ is filled by the gases which overlie the photosphere; and it is easy to see that, looked at from above, such a cavity and arrangement of the luminous filaments would present the appearances actually observed.

CHAPTER VI.

THE CHROMOSPHERE AND THE PROMINENCES.

Early Observations of Chromosphere and Prominences.—The Eclipses of 1842, 1851, and 1860.—The Eclipse of 1868.—Discovery of Janssen and Lockyer.—Arrangement of Spectroscope for Observations upon Chromosphere.—Spectrum of Chromosphere.—Lines always present. —Lines often reversed.—Motion Forms.—Double Reversal of Lines. —Distribution of Prominences.—Magnitude.—Classification of Prominences as quiescent, and eruptive or metallic.—Isolated Clouds.— Violence of Motion.—Observations of August 5, 1872.—Theories as to the Formation and Causes of the Prominences.

WHAT we see of the sun under ordinary circumstances is but a fraction of his total bulk. While by far the greater portion of the solar *mass* is included within the photosphere—the blazing cloud-layer, which seems to form the sun's true surface, and is the principal source of his light and heat—yet the larger portion of his *volume* lies without, and constitutes an atmosphere whose diameter is at least double, and its bulk therefore sevenfold that of the central globe.

Atmosphere, however, is hardly the proper term; for this outer envelope, though gaseous in the main, is not spherical, but has an outline exceedingly irregular and variable. It seems to be made up not of overlying strata of different density, but rather of flames, beams, and streamers, as transient and unstable as those of our own aurora borealis. It is divided into two portions, separated by a boundary as definite, though not so regular, as that which parts them both from the photosphere. The outer and far more extensive portion, which in texture and rarity seems to resemble the tails of comets, and may almost, without exaggeration, be likened to "the stuff that dreams are made of," is known as the "coronal atmosphere," since to it is chiefly due the "corona" or glory which surrounds the darkened sun during an eclipse, and constitutes the most impressive feature of the occasion.

At its base, and in contact with the photosphere, is what resembles a sheet of scarlet fire. The appearance, which probably indicates a fact, is as if countless jets of heated gas were issuing through vents and spiracles over the whole surface, thus clothing it with flame which heaves and tosses like the blaze of a conflagration.

This is the "chromosphere" (or chromatosphere, if one is fastidious as to the proper formation of a Greek derivative), a name first proposed by Frankland and Lockyer in 1869, and intended to signify "color-sphere," in allusion to the vivid redness of the stratum, caused by the predominance of hydrogen in these flames and clouds. It was called the "*sierra*" by Airy in 1842, and Proctor and some other writers prefer that name to the later and more common appellation.

Here and there masses of this hydrogen mixed with other substances rise to a great height, ascending far above the general level into the coronal regions, where they float like clouds, or are torn to pieces by contending currents. These cloud-masses are known as solar "prominences," or "protuberances," a non-committal sort of appellation applied in 1842, when they first attracted any considerable attention, and while it was a warmly-disputed question whether they were solar, lunar, phenomena of our own atmosphere, or even mere optical illusions. It is unfortunate that no more appropriate and graphic name has yet been found for objects of such wonderful beauty and interest.

Until recently, the solar atmosphere could be seen only at an eclipse, when the sun itself is hidden by the moon. Now, however, the spectroscope has brought the chromosphere and the prominences within the range of daily observation, so that they can be studied with nearly the same facility as the spots and faculæ, and a fresh field of great interest and importance is thus opened to science.

It seems hardly possible that the ancients should have failed to notice, even with the naked eye, in some one of the many eclipses on record, the presence of blazing, star-like objects around the edge of the moon, but we find no mention of any thing of the kind, although the corona is described as we see it now. On this ground some have surmised that the sun has really undergone a change in modern times, and that the chromosphere and prominences are a new development in the solar history. But such mere negative evidence is altogether insufficient as a foundation for so important a conclusion.

The earliest recorded observation of the prominences is probably that of Vassenius, a Swedish astronomer, who, during the total eclipse of 1733, noticed three or four small pinkish clouds, entirely detached from the limb of the moon, and, as he supposed, floating in the lunar atmosphere. At that time this was the most natural interpretation of the appearance, since the fact that the moon has no atmosphere was not yet ascertained.

The Spanish admiral, Don Ulloa, in his account of

the eclipse of 1778, describes a point of red light which made its appearance on the western limb of the moon about a minute and a quarter before the emergence of the sun. At first small and faint, it grew brighter and brighter until extinguished by the returning sunlight. He supposed that the phenomenon was caused by a hole or fissure in the body of the moon; but, with our present knowledge, there can be little doubt that it was simply a prominence gradually uncovered by her motion.

The chromosphere seems to have been seen even earlier than the prominences: thus Captain Stannyan, in a report on the eclipse of 1706, observed by him at Berne, noticed that the emersion of the sun was preceded by a blood-red streak of light, visible for six or seven seconds upon the western limb. Halley and Louville saw the same thing in 1715. Halley says that two or three seconds before the emersion a long and very narrow streak of a dusky but strong red light seemed to color the dark edge of the moon on the western edge where the sun was about to reappear. Louville's account agrees substantially with this, and he further describes the precautions he used to satisfy himself that the phenomenon was no mere optical illusion, nor due to any imperfection of his telescope.

In eclipses that followed that of 1733, the chromosphere and prominences seem to have attracted but little attention, even if they were observed at all. Something of the sort appears to have been noticed by Ferrers in 1806, but the main interest of his observation lay in a different direction.

In July, 1842, a great eclipse occurred, and the shadow of the moon described a wide belt running across southern France, northern Italy, and a portion of Austria. The eclipse was carefully observed by many of the most noted astronomers of the world, and so completely had previous observations of the kind been forgotten, that the prominences, which appeared then with great brilliance, were regarded with extreme surprise, and became objects of warm discussion, not only as to their cause and location, but even as to their very existence. Some thought them mountains upon the sun, some that they were solar flames, and others, clouds floating in the sun's atmosphere. Others referred them to the moon, and yet others claimed that they were mere optical illusions. At the eclipse of 1851 (in Sweden and Norway), similar observations were repeated, and, as a result of the discussions and comparison of observations which followed, astronomers generally became satisfied that the prominences are real phenomena of the solar atmosphere, in many respects analogous to our terrestrial clouds; and several came more or less confidently to the conclusion, now known to be true (see Grant's "History of Physical Astronomy"), that the sun is entirely surrounded with a continuous stratum of the same substance. Many, however, remained unconvinced: Faye, for instance, still asserted them to be mere optical illusions, or mirages.

In the eclipse of 1860, photography was for the first time employed on such an occasion with anything like success. The results of Secchi and De La Rue removed all remaining doubts as to the real existence and solar character of the objects in question, by exhibiting them upon their plates gradually covered on one side and uncovered on the other side of the sun by the progress of the moon.

Secchi thus sums up his conclusions, which have been justified in almost all their details by later observations; they require few and slight corrections:

1. The prominences are not mere optical illusions; they are real phenomena pertaining to the sun. . . .

2. The prominences are collections of luminous matter of great brilliance, and possessing remarkable photographic activity. This activity is so great that many of them, which are visible in our photographs, could not be seen directly even with good instruments.

3. Some protuberances float entirely free in the solar atmosphere like clouds. If they are variable in form, their changes are so gradual as to be insensible in the space of ten minutes. (Generally, but by no means always, true.)

4. Besides the isolated and conspicuous protuberances there is also a layer of the same luminous substance which surrounds the whole sun, and out of which the protuberances rise above the general level of the solar surface. . . .

5. The number of the protuberances is indefinitely great. In direct observation through the telescope the sun appeared surrounded with flames too numerous to count. . . .

6. The height of the protuberances is very great, especially when we take account of the portion hidden by the moon. One of them had a height of at least three minutes, which indicates a real altitude of more than ten times the earth's diameter. . . .

But their nature still remained a mystery; and no one could well be blamed for thinking it must always remain so to some degree. At that time it could hardly be hoped that we should ever be able to ascertain their chemical constitution, and measure the velocities of their motions. And yet this has been done. Before the great Indian eclipse of August 18, 1868, the spectroscope had been invented (it was, indeed, already in its infancy in 1860), and applied to astronomical research with the most astonishing and important results.

Every one is more or less familiar with the story of this eclipse. Herschel, Tennant, Pogson, Rayet, and Janssen, all made substantially the same report. They found the spectrum of the prominences observed to consist of bright lines, and conspicuous among them were the lines of hydrogen. There were some serious discrepancies, indeed, among their observations, not only as to the number of the bright lines seen, which is not to be wondered at, but as to their position. Thus, Rayet (who saw more lines than any one else) identified the red line observed with B instead of C; and all the observers mistook the yellow line they saw for that of sodium.

Still, their observations, taken together, completely demonstrated the fact that the prominences are enormous masses of highly-heated gaseous matter, and that hydrogen is a main constituent.

Janssen went further. The lines he saw during the eclipse were so brilliant that he felt sure he could see them again in the full sunlight. He was prevented by clouds from trying the experiment the same afternoon, after the close of the eclipse; but the next morning the sun rose unobscured, and, as soon as he had completed the necessary adjustments, and directed his instrument to the portion of the sun's limb where the day before the most brilliant prominence appeared, the same lines came out again, clear and bright; and now, of course, there was no difficulty in determining at leisure, and with almost absolute accuracy, their position in the spectrum. He immediately confirmed his first conclusion, that hydrogen is the most conspicuous component of the prominences, but found that the yellow line must be referred to some different element than sodium, being somewhat more refrangible then the D lines.

He found also that, by slightly moving his telescope and causing the image of the sun's limb to take different positions with reference to the slit of his spectroscope, he could even trace out the form and measure the dimensions of the prominences; and he remained at his station for several days, engaged in these novel and exceedingly interesting observations.

Of course, he immediately sent home a report of his eclipse-work, and of his new discovery, but, as his station at Guntoor, in eastern India, was farther from mail communication with Europe than those upon the western coast of the peninsula, his letter did not reach France until some week or two after the accounts of the other observers; when it did arrive, it came to Paris, in company with a communication from Mr. Lockyer, announcing the same discovery, made independently, and even more creditably, since with Mr. Lockyer it was not suggested by anything he had seen, but was thought out from fundamental principles.

Nearly two years previously the idea had occurred to him (and, indeed, to others also, though he was the first to publish it) that, if the protuberances are gaseous, so as to give a spectrum of bright lines, those lines ought to be visible in a spectroscope of sufficient power, even in broad daylight. The principle is simply this:

Under ordinary circumstances the protuberances are invisible, for the same reason as the stars in the daytime: they are hidden by the intense light reflected from the particles of our own atmosphere near the sun's place in the sky, and, if we could only sufficiently weaken this aërial illumination, without at the same time weakening *their* light, the end would be gained.

And the spectroscope accomplishes precisely this very thing. Since the air-light is reflected sunshine, it of course presents the same spectrum as sunlight, a continuous band of color crossed by dark lines. Now, this sort of spectrum is greatly weakened by every increase of dispersive power, because the light is spread out into a longer ribbon and made to cover a more extended area. On the other hand, a spectrum of bright lines undergoes no such weakening by an increase in the dispersive power of the spectroscope. The bright lines are only more widely separated—not in the least diffused or shorn of their brightness. If, then, the image of the sun, formed by a telescope, be examined with a spectroscope, one might hope to see at the edge of the disk the bright lines belonging to the spectrum of the prominences, in case they are really gaseous.

Mr. Lockyer and Mr. Huggins both tried the experiment as early as 1867, but without success; partly because their instruments had not sufficient power to bring out the lines conspicuously, but more because they did not know whereabouts in the spectrum to look for them, and were not even sure of their existence. At any rate, as soon as the discovery was announced, Mr. Huggins immediately saw the lines without difficulty, with the same instrument which had failed to show them to him before. It is a fact, too often forgotten, that to perceive a thing known to exist does not require one half the instrumental power or acuteness of sense as to discover it.

Mr. Lockyer, immediately after his suggestion was published, had set about procuring a suitable instrument, and was assisted by a grant from the treasury of the Royal Society. After a long delay, consequent in part upon the death of the optician who had first undertaken its construction, and partly due to other causes, he received the new spectroscope just as the report of Herschel's and Tennant's observations reached England. Hastily adjusting the instrument, not yet entirely completed, he at once applied it to his telescope, and without difficulty found the lines, and verified their position. He immediately also discovered them to be visible around the whole circumference of the sun, and consequently that the protuberances are mere extensions of a continuous solar envelope, to which, as mentioned above, was given the name of Chromosphere. (He does not seem to have been aware of the earlier and similar conclusions of Arago, Grant, Secchi, and others.) He at once communicated his results to the Royal Society, and also to the French Academy of Sciences, and, by one of the curious coincidences which so frequently occur, his letter and Janssen's were read at the same meeting, and within a few minutes of each other.

The discovery excited the greatest enthusiasm, and in 1872 the French Government struck a gold medal in honor of the two astronomers, bearing their united effigies.

It immediately occurred to several observers, Janssen, Lockyer, Zöllner, and others, that by giving a rapid motion of vibration or rotation to the slit of the spectroscope it would be possible to perceive the whole contour and detail of a protuberance at once, but it seems to have been reserved for Mr. Huggins to be the first to show practically that a still simpler device would answer the same purpose. With a spectroscope of sufficient dispersive power it is only necessary to widen the slit of the instrument by the proper adjusting screw. As the slit is widened, more and more of the protuberance becomes visible, and, if not too large, the whole can be seen at once: with the widening of the slit, however, the brightness of the background increases, so that the finer details of the object are less clearly seen, and a limit is soon reached beyond which further widening is disadvantageous. The higher the dispersive power of the spectroscope the wider the slit that can be used, and the larger the protuberance that can be examined as a whole.

Fig. 40.

Huggins's First Observation of a Prominence in Full Sunshine.

Mr. Huggins's first successful observation of the form of a solar protuberance was made on February 13, 1869. Fig. 40, copied from the "Proceedings of the Royal Society," presents his delineation of what he saw. As his instrument had only the dispersive power of two prisms, and included in its field of view a large portion of the spectrum at once, he found it necessary to supplement its powers by using a red glass to cut off stray light of other colors, and by inserting a diaphragm at the focus of the small telescope of the spectroscope to limit the field of view to the portion of the spectrum immediately adjoining the C line. With the instruments now in use, these precautions are seldom necessary.

It may be noticed, in passing, that Mr. Huggins had previously (and has subsequently) made many experiments with different absorbing media, in hopes of finding some substance which, by cutting off all light of other color than that emitted by the prominences, should render them visible in the telescope; thus far, however, without success.

FIG. 41.

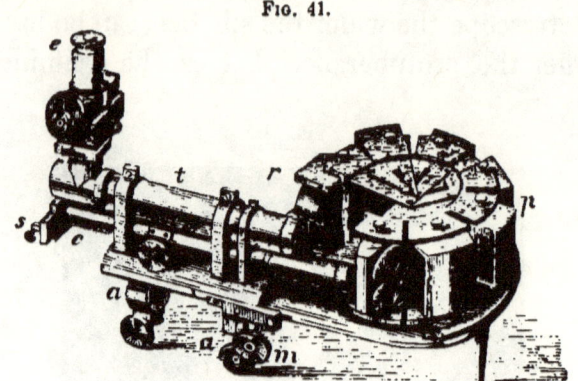

SPECTROSCOPE, WITH TRAIN OF PRISMS.

The spectroscopes used by different astronomers for observations of this sort differ greatly in form and power. Fig. 41 represents the one employed at the Shattuck Observatory of Dartmouth College, and several of our American observatories are supplied with instruments similarly arranged. The light passes from the collimator c, through the train of prisms p, near their bases, and, by two reflections in a rectangular prism, r, is transferred to the upper story, so to speak, of the prism-train, and made to return to the telescope t, finally reaching the eye at e. It thus twice traverses a train of six prisms, and the dispersive power of the instrument is twelve times as great as it would be with only one prism. The diameter of the collimator is a little less than an inch, and its length ten inches. The whole instrument, powerful as it is, only weighs about fourteen pounds, and occupies a space of about 15 in.

× 6 in. × 5 in. It is also *automatic*, i. e., the tangent screw *m* keeps the train of prisms adjusted to their position of minimum deviation by the same movement which brings the different portions of the spectrum to the center of the field of view, and the milled head *f* focuses both the collimator and telescope simultaneously.

The spectroscope is attached to the equatorial telescope, to which it belongs, by means of the clamping rings *a, a*. These slide upon a stout metal rod, firmly fastened to the telescope in such a way that the slit *s*, of the instrument, can be placed exactly at the focus of the object-glass, where the image of the sun is formed. This instrument, attached to the telescope, has already been figured upon page 78.

Instruments in which the prism-train is replaced by a diffraction-grating are still more powerful; and more convenient also, since the observer has the great advantage of being able to select, within certain limits, the amount of dispersion best suited to his purpose by simply turning the grating so as to utilize the different orders of spectra—an operation easier and more rapid than that of rearranging the prism-train. Diffraction spectroscopes have, however, one slight disadvantage. When used with the open slit, the forms of objects seen through the slit are somewhat distorted, being either compressed or extended in a direction at right angles to the slit. When the grating is so placed that the inclination of its surface to the view-telescope is greater than to the collimator (as in the figure on page 74), compression occurs. In this case, the edge of the slit being placed tangential to the sun's limb, as is usual, prominences on the edge of the sun appear to have their height reduced. Of course, the reverse takes place when the grating is placed the opposite way. This distortion, however, is of little importance, as its amount is easily calculated and allowed for when necessary.* A similar distortion is produced by prismatic spectroscopes when the prisms are not adjusted strictly to their position of minimum deviation.

The diffraction instrument, which the writer is accustomed to use for solar observations at Princeton, has already been figured on page 75.

With a telescope of not less than four inches aperture, equatorially mounted, and a spectroscope of dispersive power not less than that of five or six ordinary prisms, the observer is equipped for the study of the chromosphere and prominences. He may either study the spectrum as such, using the instrument with a narrow slit, or he may employ it with widened slit simply as a means of viewing the prominences and studying their forms and changes.

The spectra of the chromosphere and prominences are very interesting in their relations to that of the photosphere, and present many peculiarities which are not yet fully explained. At times and in places where some special disturbance is going on—frequently in the neighborhood of spots at the times when they are just passing around the limb of the disk—the spectrum, at the base of the chromosphere, is very complicated, consisting of hundreds of bright lines. In the course of a few weeks of observation at Sherman in 1872, the writer made out a list of two hundred and seventy-three, and

---

* The formula for the calculation is simply $H = h\dfrac{\sin. l}{\sin. k}$, in which H is the true height of the object seen through the slit; h is its apparent height, and k and l are the inclinations of the surface of the grating to the collimator and view-telescope respectively.

more recent observations with the Princeton spectroscope show that the real number must be vastly greater; perhaps it may be fully doubled by a little watchfulness. The majority of the lines, however, are seen only occasionally, for a few minutes at a time, when the gases and vapors, which generally lie low, mainly in the interstices of the clouds which constitute the photosphere, and below its upper surface, are elevated for the time being by some eruptive action. For the most part, the lines which appear only at such times are simply "reversals" of the more prominent dark lines of the ordinary solar spectrum. But the selection of the lines seems most capricious; one is taken, and another left, though belonging to the same element, of equal intensity, and close beside the first. It is evident that the subject needs a detailed and careful study, combining solar observations with laboratory-work upon the spectra of the elements concerned, before a satisfactory account can be given of all the peculiar behavior observed.

The lines composing the true chromosphere spectrum, if we may call it so (that is, those which are always observable in it with suitable appliances), are not very numerous, and we give the following list, designating them by their wave-length, as given by Ångström:

1. 7055 ±. Element unknown.
2. 6561·8, C. Hydrogen (Hα).
3. 5874·9, D$_3$. Unknown element:—Frankland's "*helium.*"
4. 5315·9. The corona-line, element unknown.
5. 4860·6, F. Hydrogen (Hβ).
6. 4471·2, f. Cerium?
7. 4340·1, near G. Hydrogen (Hγ).
8. 4101·2, h. Hydrogen (Hδ).
9. 3969 ? Element unknown.
10. 3967·9, H. Hydrogen, probably.
11. 3932·8, K or H$_2$. Hydrogen, probably.

The first line is generally very difficult to see, though sometimes pretty conspicuous. It is in the red, between B and a, and has a very faint corresponding dark line. No. 3 has no dark line corresponding as a usual thing, though occasionally one appears, especially in the neighborhood of sun-spots. No. 9 is quite within the broad shade of the H-line, which thus appears double in the chromosphere spectrum. The line, however, does not probably belong to the same element as H, because in the sun-spot spectrum H appears bright but *single*, as has been already mentioned in another place.

The eleven lines mentioned above are invariably present in the spectrum of the chromosphere; a much larger number make their appearance on very slight provocation. They are:

1′.	6676·9.	Iron.	17′. 4933·4.	Barium.
2′.	6429·9.	?	18′. 4923·1.	Iron.
3′.	6140·6.	Barium.	19′. 4921·3.	?
4′.	5895·0, D_1.	Sodium.	20′. 4918·2.	Iron.
5′.	5889·0, D_2.	"	21′. 4899·3.	Barium.
6′.	5361·9.	Iron.	22′. 4500·3.	Titanium.
7′.	5283·4.	?	23′. 4490·9.	Manganese.
8′.	5275·0.	?	24′. 4489·4.	Manganese and iron.
9′.	5233·6.	Manganese.	25′. 4468·5.	Titanium.
10′.	5197·0.	?	26′. 4394·6.	?
11′.	5183·0, b_1.	Magnesium.	27′. 4245·2.	Iron.
12′.	5172·0, b_2.	"	28′. 4235·5.	"
13′.	5168·3, b_3.	Iron and nickel.	29′. 4233·0.	Iron and calcium.
14′.	5166·7, b_4.	Magnesium.	30′. 4215·0.	Calcium and strontium.
15′.	5017·6.	Iron and nickel.		
16′.	5015·0.	?	31′. 4077·0.	Calcium.

It is not intended, however, to intimate that, if one of these appears, all of them will do so, nor that they are equally conspicuous or equally common. To a certain degree also, their selection by the writer is arbitrary, for there are nearly as many more which are seen pretty frequently, and some of them may very possibly be found hereafter to deserve a place upon the list rather than some that have been included.

It requires careful manipulation to bring out the fainter and finer lines satisfactorily. The slit must be adjusted with extreme care to the focal plane of the rays under examination, placed tangential to the solar image, and brought exactly to the edge of the disk. A thousandth of an inch in its position will often make the whole difference between a successful observation and its failure, and even a slight unsteadiness of the air will diminish the number of bright lines visible by at least one half.

As the majority of the lines are only developed by more or less unusual disturbances of the solar surface, it naturally happens that one very often finds them distorted or displaced by the motions of the gases along the line of sight (toward or from the observer), as explained in a previous chapter, producing what Lockyer calls "motion-forms." Occasionally, also, we meet with "double reversals," so called, especially in the lines of magnesium and sodium. The (dark) lines of these substances are rather wide in the solar spectrum. When reversed in the chromosphere spectrum, the phenomenon usually consists of a thin bright line down the center of the wider dark band: in a double reversal the bright line widens and a fine dark line appears in *its* center, so that we have a central dark line, a bright one on each side of it, and outside of the bright lines a dark shade on both sides. Fig. 42 represents such a double reversal of the D-lines observed by the writer on several occasions in 1880. The phenomenon seems to be due to the presence of an unusual quantity of the vapor at a considerable density, and is the precise correlative of what is sometimes seen in the spectrum of a sodium-flame. The two D-lines of sodium each becomes itself

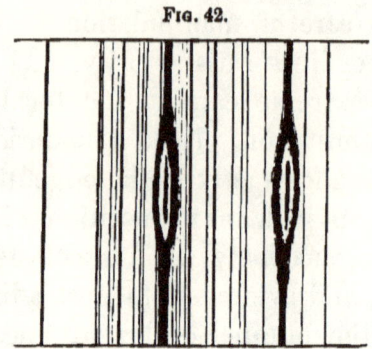

FIG. 42.

DOUBLE-REVERSAL OF THE D-LINES.—(October, 1880.)

double, so that we get pairs of bright lines in place of single lines. The electric arc often shows this still more finely.

Generally speaking, the spectrum of a prominence is simpler than that of the chromosphere at its base. We seldom find any lines except C, D_3, F, Hγ, and h, at a considerable elevation above the photosphere, though H, K, and f are sometimes met with. On rare occasions, also, the vapors of sodium and magnesium are carried into the higher regions, and once or twice the writer has seen the line No. 1 of the second list (6676·9) in the upper portions of a prominence.

When the spectroscope is used as a means of rendering visible the forms and features of the prominences, the only difference is that the slit is more or less widened.

The telescope is directed so that the solar image shall fall with that portion of its limb which is to be examined just tangent to the opened slit, as in Fig. 43, which represents the slit-plate of the spectroscope of its actual size, with the image of the sun in position for observation.

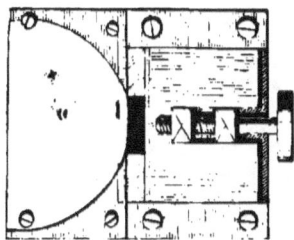

FIG. 43.

OPENED SLIT OF THE SPECTROSCOPE.

If, now, a prominence exists at this part of the sun's limb (as would probably be the case, considering the proximity of the spot shown in the figure), and if the spectroscope itself is so adjusted that the C-line falls in the center of the field of view, then, on looking into the eye-piece, one will see something much like Fig. 44. The red portion of the spectrum will stretch athwart the field of view like a scarlet ribbon, with a darkish band across it, and in that band will appear the prominences, like scarlet clouds—so like our own terrestrial clouds, indeed, in form and texture, that the resemblance is quite startling: one might almost think he was looking out through a partly-opened door upon a sunset sky, except that there is no variety or contrast of color; all the cloudlets are of the same pure scarlet hue. Along the edge of the opening is seen the chromosphere, more brilliant than the clouds which rise from it or float above it, and for the most part made up of minute tongues and filaments. Usually, however, the definition of the chromosphere is less distinct than that of the higher clouds. The reason is, that close to the limb of the sun, where the temperature and pressure are highest, the hydrogen is in such a state that the lines of its spectrum are widened and "winged," something like those of magnesium, though to a less extent. Each point in the chromosphere, therefore, when viewed through the opened slit, appears not as a *point*, but as a *short line*, directed lengthwise in the spectrum. As the length of this line depends upon the dispersive power of the spectroscope, it is easy to see that it is possible to go too far in this respect. The lower the dispersion the more *distinct* the image obtained, but also the fainter as compared with the background upon which it is seen.

FIG. 44.

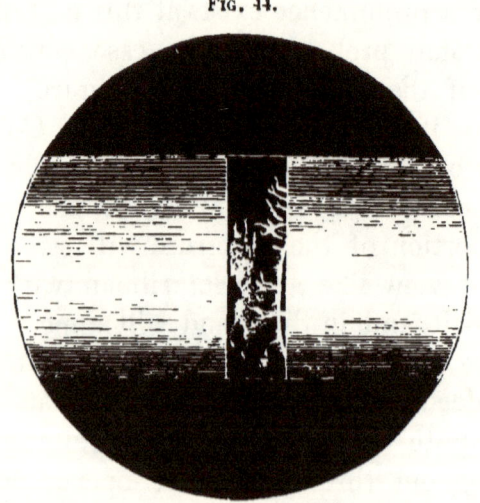

CHROMOSPHERE AND PROMINENCES AS SEEN IN THE SPECTRUM.

If the spectroscope is adjusted upon the F-line, instead of C, then a similar image of the prominences and chromosphere is seen, only blue instead of scarlet; usually, however, since the F-line is hazier and more winged than C, this blue image is somewhat less perfect in its details and definition, and is therefore less used for observation. Similar effects are obtained by means of the yellow line near D, and the violet line near G. By setting the spectroscope upon this latter line and attaching a small camera to the eye-piece, it is even possible to photograph a bright protuberance; but the light is so feeble, the image so small, the time of exposure needed so long, and the requisite accuracy of motion in the clock-work which drives the telescope so difficult of attainment, that thus far no pictures of any real value have been obtained in this manner.

Professor Winlock and Mr. Lockyer have attempted, by using an annular opening instead of the ordinary slit, to obtain a view of the whole circumference of the sun at once, and have succeeded. With a spectroscope of sufficient power, and adjustments delicate enough, the thing can be done; but as yet no very satisfactory results appear to have been reached. We are still obliged to examine the circumference of the sun piecemeal, so to speak, readjusting the instrument at each point, to make the slit tangential to the limb.

The number of protuberances of considerable magnitude (exceeding ten thousand miles in altitude), visible at any one time on the circumference of the sun, is never very great, rarely reaching twenty-five or thirty. Their number, however, varies extremely with the number of sun-spots: during the late sun-spot minimum in 1878–'79, there were not unfrequently occasions when not a single one could be found, though even during those years the more usual number was five or six— some of them of considerable size. The observations of Tacchini and Secchi have showed that their numbers closely followed the march of the sun-spots, though never falling quite so low.

Their distribution on the sun's surface is in some respects similar to that of the spots, but with important

200 THE SUN.

differences. The spots are confined within 40° of the sun's equator, being most numerous at a solar latitude of about 20° on each hemisphere. Now, the protuberances are most numerous precisely where the spots are most abundant, but they do not disappear at a latitude of 40°; they are found even at the poles, and from the latitude of 60° actually increase in number to a latitude of about 75°.

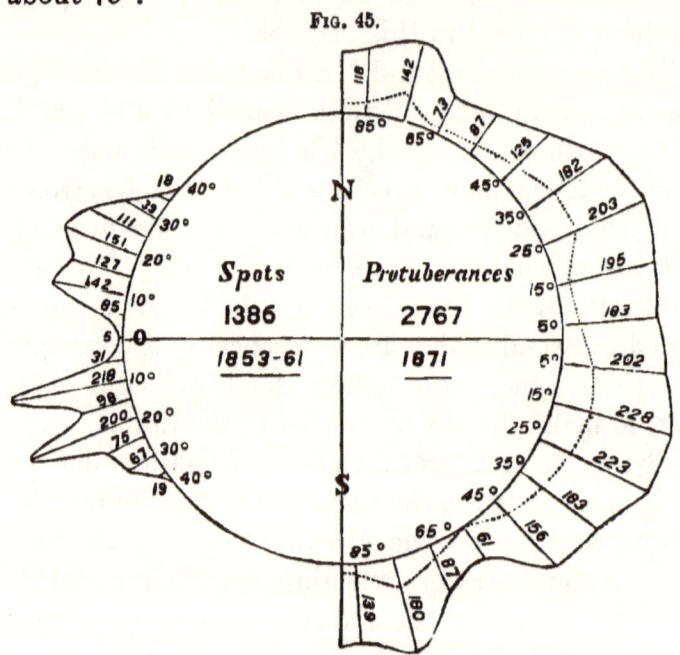

RELATIVE FREQUENCY OF PROTUBERANCES AND SUN-SPOTS.

The annexed diagram, Fig. 45, represents the relative frequency of the protuberances and spots on the different portions of the solar surface. On the left side is given the result of Carrington's observation of 1,386 spots between 1853 and 1861, and on the right the result of Secchi's observations of 2,767 * protuberances in

* The 2,767 prominences are not all different ones. If any of the prominences observed on one day remained visible the next, they were

1871. The length of each radial line represents the number of spots or protuberances observed at each particular latitude on a scale of a quarter of an inch to the hundred; for example, Secchi gives 228 protuberances as the number observed during the period of his work between 10° and 20° of south latitude, and the corresponding line drawn at 15° south, on the left-hand side of the figure, is therefore made $\frac{228}{400}$ or ·57 of an inch long. The other lines are laid off in the same way, and thus the irregular curve drawn through their extremities represents to the eye the relative frequency of these phenomena in the different solar latitudes. The dotted line on the right-hand side represents in the same manner and on the same scale the distribution of the larger protuberances, having an altitude of more than 1', or 27,000 miles.

A mere inspection of the diagram shows at once that, while the prominences may, and in fact often do, have a close connection with the spots, they are yet to some extent independent phenomena.

A careful study of the subject shows that they are much more closely related to the faculæ. In many cases at least, faculæ, when followed to the limb of the sun, have been found to be surrounded by prominences, and there is reason to suppose that the fact is a general one. The spots, on the other hand, when they reach the border of the sun's image, are commonly surrounded by prominences more or less completely, but seldom overlaid by them. Indeed, Respighi asserts (and the

recorded afresh; and, as a prominence near the pole would be carried but slowly out of sight by the sun's rotation, it is thus easy to see how the number of prominences recorded in the polar regions is so large, notwithstanding the smaller area of each zone of 5° width, as compared with a similar zone near the equator.

most careful observations we have been able to make confirm his statement) that as a general rule the chromosphere is considerably depressed immediately over a spot. Secchi, however, denies this.

The protuberances differ greatly in magnitude. The average depth of the chromosphere is not far from 10″ or 12″, or about 5,000 or 6,000 miles, and it is not, therefore, customary to note as a prominence any cloud with an elevation of less than 15″ or 20″—7,000 to 9,000 miles. Of the 2,767 already quoted, 1,964 attained an altitude of 40″, or 18,000 miles, and it is worthy of notice that the smaller ones are so few, only about one third of the whole: 751, or nearly one fourth of the whole, reached a height of over 1′, or 28,000 miles; the precise number which reached greater elevations is not mentioned, but several exceeded 3′, or 84,000 miles. It is only rather rarely that they reach elevations as great as 100,000 miles. The writer has in all seen, perhaps, three or four which exceeded 150,000 miles, and Secchi has recorded one of 300,000 miles. On October 7, 1880, the writer observed one which attained the hitherto unprecedented height of over 13′ of arc, or 350,000 miles. When first seen, on the southeast limb of the sun, about 10.30 A. M., it was a "horn" of ordinary appearance, some 40,000 miles in elevation, and attracted no special attention. When next seen, half an hour later, it had become very brilliant and had doubled its height: during the next hour it stretched upward until it reached the enormous altitude mentioned, breaking up into filaments which gradually faded away, until, by 12.30 P. M., there was nothing left. A telescopic examination of the sun's disk showed nothing to account for such an extraordinary outburst, except some small and not very brilliant faculæ. While it was extending upward most rapidly a violent cyclonic motion was shown in the lower part by the displacement of the spectrum-lines.

In their form and structure the protuberances differ as widely as in their magnitude. Two principal classes

ERUPTIVE PROMINENCES.

Three figures, of the same prominence, seen July 25, 1872.
FIG. 46.

As seen at 2.15 p. m.

FIG. 49.

SPIKES.

FIG. 47.

As seen at 2.45 p. m.

FIG. 50.

SHEAF AND VOLUTES.

FIG. 48.

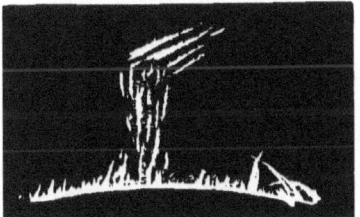

As seen at 3.30 p. m.
100,000 miles to the inch.

FIG. 51.

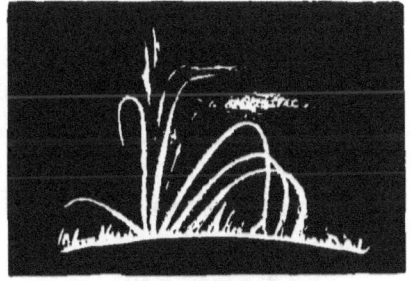

JETS.

are recognized by all observers—the *quiescent, cloud-formed*, or hydrogenous, and the *eruptive* or metallic. By Secchi these are each further subdivided into several sub-classes or varieties, between which, however, it is not always easy to maintain the distinctions.

The quiescent prominences in form and texture resemble, with almost perfect exactness, our terrestrial clouds, and differ among themselves as much and in the same manner. The familiar cirrus and stratus types are very common, the former especially, while the cumulus and cumulo-stratus are less frequent. The protuberances of this class are often of enormous magnitude, especially in their horizontal extent (but the highest elevations are attained by those of the eruptive order), and are comparatively permanent, remaining often for hours and days without serious change; near the poles they sometimes persist through a whole solar revolution of twenty-seven days. Sometimes they appear to lie upon the limb of the sun like a bank of clouds in the horizon; probably because they are so far from the edge of the disk that only their upper portions are in sight. When seen in their full extent they are ordinarily connected to the underlying chromosphere by slender columns, which are usually smallest at the base, and appear often to be made up of separate filaments closely intertwined, and expanding upward. Sometimes the whole under surface is fringed with down-hanging filaments, which remind one of a summer shower falling from a heavy thunder-cloud. Sometimes they float entirely free from the chromosphere; indeed, as a general rule, the layer clouds are attended by detached cloudlets for the most part horizontal in their arrangement.

The figures give an idea of some of the general

THE CHROMOSPHERE AND THE PROMINENCES. 205

QUIESCENT PROMINENCES.

Scale, 75,000 miles to the inch.

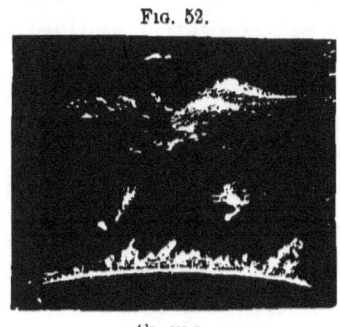

Fig. 52.

CLOUDS.

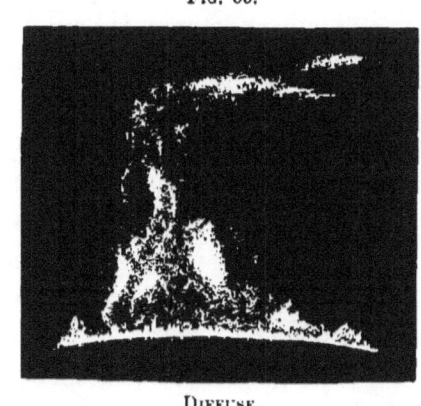

Fig. 55.

DIFFUSE.

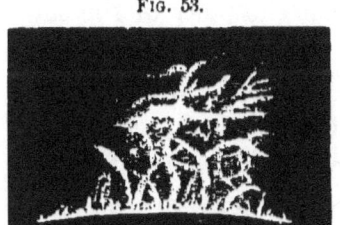

Fig. 53.

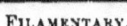

FILAMENTARY.

Fig. 56.

STEMMED.

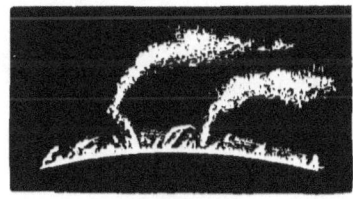

Fig. 54.

PLUMES.

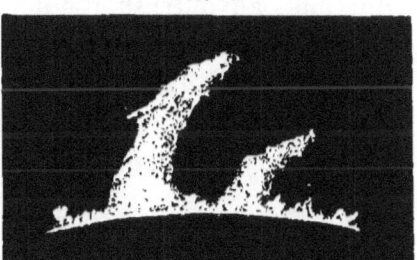

Fig. 57.

HORNS.

appearances of this class of prominences, but their delicate, filmy beauty can be adequately rendered only by a far more elaborate style of engraving.

Their spectrum is usually very simple, consisting of the four lines of hydrogen and the orange D'—hence the appellation hydrogenous. Occasionally the sodium and magnesium lines also appear, and that even near the summit of the clouds; and this phenomenon was so much more frequently observed in the clear atmosphere of Sherman as to suggest that, if the power of our spectroscopes were sufficiently increased, it would cease to be unusual.

The genesis of this sort of prominence is problematical. They have been commonly looked upon as the *débris* and relics of eruptions, consisting of gases which have been ejected from beneath the solar surface, and then abandoned to the action of the currents of the sun's upper atmosphere. But near the poles of the sun distinctively eruptive prominences never appear, and there is no evidence of aërial currents which would transport to those regions matters ejected nearer the sun's equator. Indeed, the whole appearance of these objects indicates that they originate where we see them. Possibly, although in the polar regions there are no violent eruptions, there yet may be a quiet outpouring of heated hydrogen sufficient to account for their production—an outrush issuing through the smaller pores of the solar surface, which abound near the poles as well as elsewhere.

But Secchi reports an observation which, if correct, puts a very different face upon the matter.* He has

---

* Until very recently no other spectroscopist has confirmed this observation. On October 13, 1880, the writer for the first time met with the same phenomenon. A small, bright cloud appeared on that day, about

seen isolated cloudlets form and grow spontaneously without any perceptible connection with the chromosphere or other masses of hydrogen, just as in our own atmosphere clouds form from aqueous vapor, already present in the air, but invisible until some local cooling or change of pressure causes its condensation. These prominences are, therefore, formed by some local heating or other luminous excitement of hydrogen already present, and not by any transportation and aggregation of materials from a distance. The precise nature of the action which produces this effect it would not be possible to assign at present; but it is worthy of note that the observations of the eclipse of 1871, by Lockyer and others, rather favor this view, by showing that hydrogen, in a feebly luminous condition, is found all around the sun, and at a very great altitude—far above the ordinary range of prominences.

The eruptive prominences are very different, consisting usually of brilliant spikes or jets, which change their form and brightness very rapidly. For the most part they attain altitudes of not more than 20,000 or 30,000 miles, but occasionally they rise far higher than even the largest of the clouds of the preceding class. Their spectrum is very complicated, especially near their base, and often filled with bright lines, those of sodium, magnesium, barium, iron, and titanium, being especially conspicuous, while calcium, chromium, manganese, and probably sulphur, are by no means rare,

11 A. M., at an elevation of some $2\frac{1}{2}'$ (67,500 miles) above the limb, without any evident cause or any visible connection with the chromosphere below. It grew rapidly without any sensible rising or falling, and in an hour developed into a large stratiform cloud, irregular on the upper surface, but nearly flat beneath. From this lower surface pendent filaments grew out, and by the middle of the afternoon the object had become one of the ordinary stemmed prominences, much like Fig. 56.

208   THE SUN.

Scale, 75,000 miles to the inch.

FIG. 58.

VERTICAL FILAMENTS.

FIG. 59.

CYCLONE.

FIG. 60.

FLAMES.

FIG. 61.

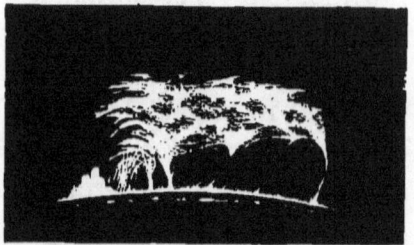

PROMINENCE AS IT APPEARED AT HALF-PAST TWELVE O'CLOCK, SEPTEMBER 7, 1871.

FIG. 62.

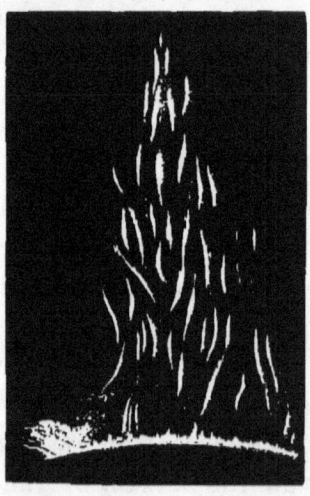

AS THE ABOVE APPEARED HALF AN HOUR LATER, WHEN THE UP-RUSHING HYDROGEN ATTAINED A HEIGHT OF MORE THAN 200,000 MILES.

FIG. 63.

SPOT NEAR THE SUN'S LIMB, WITH ACCOMPANYING JETS OF HYDROGEN, AS SEEN OCTOBER 5, 1871.

and for this reason Secchi calls them *metallic* prominences.

They usually appear in the immediate neighborhood of a spot, never occurring very near the solar poles. Their form and appearance change with great rapidity, so that the motion can almost be seen with the eye—an interval of fifteen or twenty minutes being often sufficient to transform, quite beyond recognition, a mass of these flames fifty thousand miles high, and sometimes embracing the whole period of their complete development or disappearance. Sometimes they consist of pointed rays, diverging in all directions, like hedgehog-spines. Sometimes they look like flames; sometimes like sheaves of grain; sometimes like whirling water-spouts, capped with a great cloud; occasionally they present most exactly the appearance of jets of liquid fire, rising and falling in graceful parabolas; frequently they carry on their edges spirals like the volutes of an Ionic column; and continually they detach filaments which rise to a great elevation, gradually expanding and growing fainter as they ascend, until the eye loses them. Our figures present some of the more common and typical forms, and illustrate their rapidity of change, but there is no end to the number of curious and interesting appearances which they exhibit under varying circumstances.

The velocity of the motions *often* exceeds a hundred miles a second, and sometimes, though very rarely, reaches two hundred miles. That we have to do with actual motions, and not with mere change of place of a luminous form, is rendered certain by the fact that the lines of the spectrum are often displaced and distorted in a manner to indicate that some of the cloud-masses are moving either toward or from the earth (and, of course, tangential to the solar surface) with similar swiftness.

Fig. 64 is a representation of a portion of the spectrum of a prominence observed at Sherman on August 3, 1872, an observation to which allusion was made in

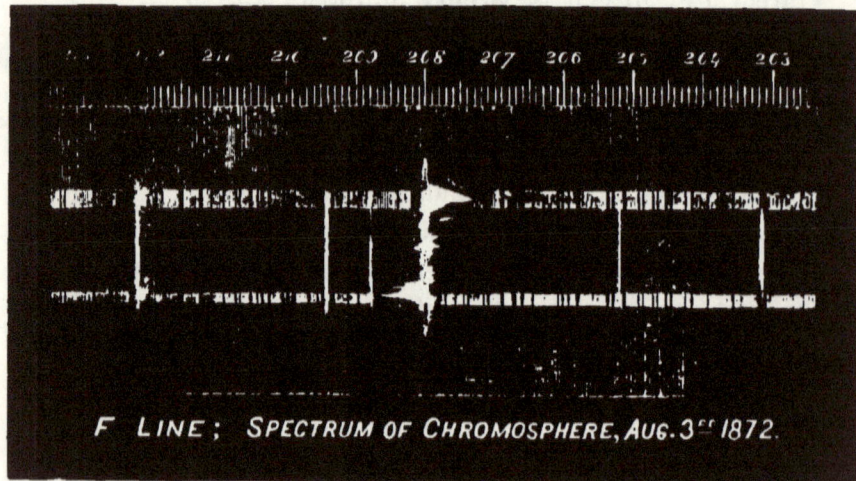

Fig. 64.

F LINE; SPECTRUM OF CHROMOSPHERE, AUG. 3rd 1872.

the preceding chapter. The F-line, at 208 of the scale, must be imagined as blazingly brilliant, and fainter bright lines appear at 203·2, 208·8, 209·4, and 212·1 (the scale is Kirchhoff's), while two bands of continuous spectrum, produced probably by the compression of the gas at the points of maximum disturbance, run the whole length of the figure. At the upper point of disturbance F is drawn out into a point reaching to 207·4 of the scale, and indicating a velocity of 230 miles a second away from us; at the lower point it extends to 208·7, and indicates a velocity of about 250 miles per second toward us. It was very noticeable that this swift motion of the hydrogen did not seem to carry with it many other substances which were at the time represented in the spectrum by their bright lines; magnesium and sodium were somewhat affected, but barium and the unknown element of the corona were not.

When we inquire what forces impart such a velocity, the subject becomes difficult. If we could admit that the surface of the sun is solid, or even liquid, as Zöllner thinks, then it would be easy to understand the phenomena as eruptions, analogous to those of volcanoes on the earth, though on the solar scale. But it is next to certain that the sun is mainly gaseous, and that its luminous surface or photosphere is a sheet of incandescent clouds, like those of the earth, except that water-droplets are replaced by droplets of the metals; and it is difficult to see how such a shell could exert sufficient confining power upon the imprisoned gases to explain such tremendous velocity in the ejected matter.

Possibly the difficulty may be met by taking account of the enormous amount of condensation which must be going on within the photosphere. To supply the heat which the sun throws off (enough to melt each minute a shell of ice nearly fifty feet thick over his entire surface) would require the condensation of enough vapor to make a sheet of liquid six feet thick in the same time —supposing, that is, the latent heat of the solar vapors not greater than that of water vapors. This, of course, is uncertain, but, so far as we know, very few if any vapors contain more latent heat than that of water, and we may therefore consider it roughly correct to estimate the continuous production of liquid as measured by the quantity named. Now, on the surface of the earth a rain-storm which deposits two inches in an hour is very uncommon—in such a storm the water falls in sheets. If we admit, then, that any considerable portion of the sun's heat is due to such a condensation of the solar vapors, it is easy to see that the quantity of liquid pouring from the solar clouds must be so enormous that the drops could not be expected to remain separate, but will almost certainly unite into more or less continuous masses or sheets, between and through which the gases ascending from beneath must make their way. And, since the weight of the vapors which ascend must continually equal that of the products of condensation which are falling, it is further evident that the upward currents, rushing through contracted channels, must move with enormous velocity, and therefore, of course, that the pressure and temperature must rapidly increase from the free surface downward. It would seem that thus we might explain how the upper surface of the hydrogen atmosphere is tormented by the up-rush from below, and how gaseous masses, thrown up from beneath, should, in the prominences, present the appearances which have been described. Nor would it be strange if veritable explosions should occur in the quasi pipes or channels through which the vapors rise, when, under the varying circumstances of pressure and temperature, the mingled gases reach their point of combination; explosions which would fairly account for such phenomena as those represented in Figs. 61 and 62, when clouds of hydrogen were thrown to an elevation of more than 200,000 miles with a velocity which *must* have exceeded at first 200 miles per second, and very probably, taking into account the resistance of the solar atmosphere, may, as Mr. Proctor has shown, have exceeded 500; a velocity sufficient to hurl a dense material entirely clear of the power of the sun's attraction, and send it out into space, never to return.

CHAPTER VII.

THE CORONA.

General Appearance of the Phenomenon.—Various Representations.—Eclipses of 1857, 1860, 1867, 1868, 1869, 1871, and 1878.—Proof that the Corona is mainly a Solar Phenomenon.—Brightness of the Corona.—Connection with Sun-Spot Period.—Spectrum of the Corona.—Application of the Analyzing and Integrating Spectroscopes.—Polarization.—Evidence of the Slitless Spectroscope as to the Constitution of the Corona.—Changes and Motions in the Corona.—Its Form and Constitution, and Theories as to its Nature and Origin.

A TOTAL eclipse of the sun is unquestionably one of the most impressive of all natural phenomena, and the corona, or aureole of light, which then surrounds the sun, is its most impressive feature. On such an occasion, if the sky is clear, the moon appears of almost inky darkness, with just sufficient illumination at the edge of the disk to bring out its rotundity in a striking manner. It looks not like a flat screen, but like a huge black ball, as it really is. From behind it stream out on all sides radiant filaments, beams, and sheets of pearly light, which reach to a distance sometimes of several degrees from the solar surface, forming an irregular stellate halo, with the black globe of the moon in its apparent center. The portion nearest the sun is of dazzling brightness, but still less brilliant than the prominences, which blaze through it like carbuncles. Generally this inner corona has a pretty uniform height, forming a ring three or four minutes of arc in width, separated by a somewhat definite outline from the outer corona, which reaches to a much greater distance, and is far more irregular in form. Usually there are several " rifts," as they have been called, like narrow beams of darkness, extending from the very edge of the sun to the outer night, and much resembling the cloud-shadows which radiate from the sun before a thunder-shower. But the edges of these rifts are frequently curved, showing them to be something else than real shadows. Sometimes there are narrow, bright streamers, as long as the rifts, or longer. These are often inclined, occasionally are even nearly tangential to the solar surface, and frequently are curved. On the whole, the corona is usually less extensive and brilliant over the solar poles, and there is a recognizable tendency to accumulations above the middle latitudes, or spot-zones; so that, speaking roughly, the corona shows a disposition to assume the form of a quadrilateral or four-rayed star, though in almost every individual case this form is greatly modified by abnormal streamers at some point or other.

Unlike the chromosphere, which seems first to have been observed, as was mentioned in the previous chapter, only a little more than a century ago, the corona has been known from antiquity, and is described by Philostratus and Plutarch in almost the same terms we should ourselves employ. And yet our knowledge of it remains very limited. The chromosphere and prominences we can now reach and study, comparatively at our leisure, by the help of the spectroscope; but the corona is still inaccessible, except during the short and precious moments of a total eclipse—in all, not more than a few days in a century—so that our knowledge of its cause and nature can grow but slowly at the best.

The character of the phenomenon is such also as to make its accurate observation exceedingly difficult; slight differences in the transparency of the atmosphere, in the sensitiveness of the observer's eye, a preoccupation of the mind by some feature which first happens to strike the attention, or a peculiarity in the manner of representing what one sees, will often make the descriptions and drawings of two observers, side by side, so discrepant that one would hardly imagine they could refer to the same object. For instance, in 1870, two naval officers on the deck of the same vessel made drawings of the corona, one of which represented it as a six-rayed star, while the other showed it as composed of two ovals crossing at right angles. In 1878 the writer, on comparing notes immediately after the eclipse with other members of his party, found that about half of them saw the corona principally extended to the east and west, while the other half, himself among them, were just as positive that it brushed out mainly to the north and south. The photographs, and other data since collected, show that the principal extension was undoubtedly along the east-and-west line, but that there were much better outlined streamers, though shorter and less brilliant, directed from the solar poles. Some eyes were more impressed by definiteness of form, others by size and luminosity.

Obviously, conclusions must be drawn from ocular impressions only with the greatest caution. Photographs are, of course, more to be trusted, as far as they go; but, even with them, a slight difference in the sensitiveness of the plate, in the exposure, or in the development, will make a great difference in the resulting picture. Neither can any photograph ever bring out everything which is visible to the eye. An exposure, sufficient to exhibit well the fainter details, will spoil the brighter features, and *vice versa*.

We can do no better than to refer one, who is curious to see how various are the representations of this wonderful object, to Mr. Ranyard's magnificent work upon the observations made during total solar eclipses, published as Volume XLI of the "Memoirs of the Royal Astronomical Society of Great Britain." In it he has reproduced nearly a hundred different drawings and photographs of the corona, as seen during the eclipses since 1850. The steel engravings of the eclipses of

Fig. 65.

Corona as observed by Liais in 1857.

1870 and 1871, based upon the photographs then made, are by far the most accurate and beautiful representations of the corona anywhere to be found. We have copied a few of his woodcuts, which give an idea of the more remarkable features of the phenomenon, and

FIG. 66.

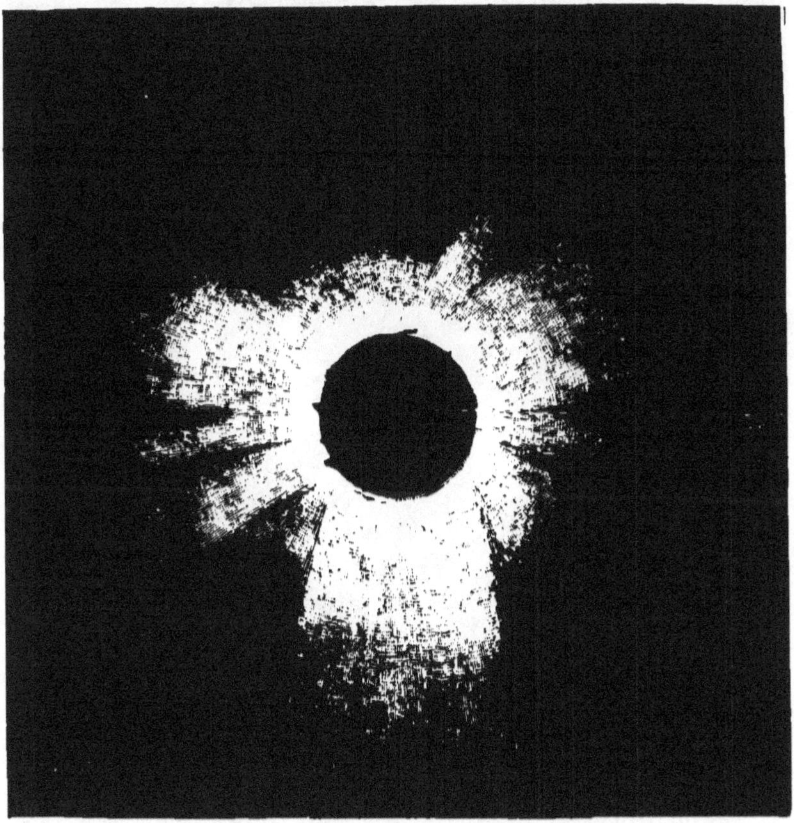

CORONA OF 1860.—SECCHI.

exhibit the differences between its character and appearance on different occasions; we have added also a picture of the corona as seen in 1878, in which we have combined the sketches of several observers with our own impressions. Woodcuts, however, are not competent to bring out the peculiar filmy, nebulous character of many of the details, which can be fairly represented only by steel engraving.

Fig. 67.

Corona of 1860.—Tempel.

The drawing of Liais, Fig. 65, shows the "petal"-like forms which have been noticed in the corona at other times, but seem to have been especially prominent in the eclipse of 1857. The figures of the corona of 1860, by Secchi and Tempel (Figs. 66, 67), show how widely observers only a few miles apart will differ in their impressions.

THE CORONA. 219

The drawing of Grosch in 1867 (Fig. 68) is interesting in comparison with that of 1878, as showing the state of the corona at two similar times of sun-spot minimum. The long extensions of faint illumination in the direction of the sun's equator and the short but vivid brushes in the polar regions are notable in both.

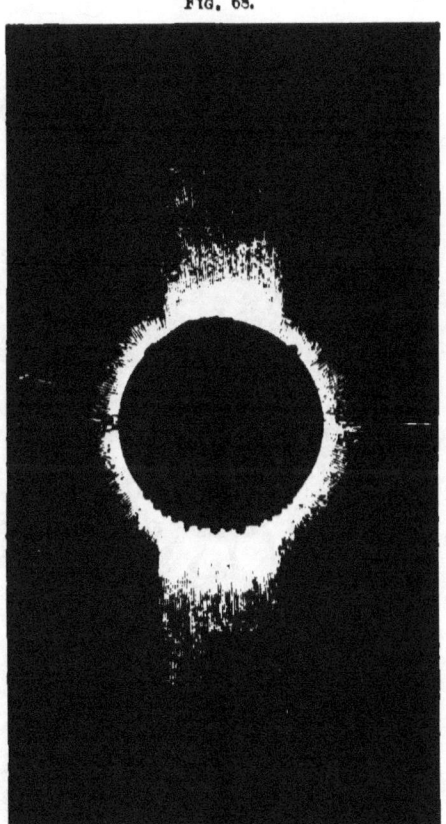

FIG. 68.

CORONA OF 1867.—GROSCH.

Bullock's picture of the eclipse of 1868 (Fig. 69) shows a larger and more irregular corona than usual. The drawing of Schott (Fig. 70), on the other hand, shows the corona of 1869 much smaller and more brilliant than ordinary, and the writer can vouch for it as giving pretty accurately the impression which he himself received at the time.

Many of our readers, no doubt, have seen a much more impressive picture of the same corona, made by Mr. Gilman at Sioux City, and published in the eclipse report of the United States Naval Observatory (repro-

Fig. 69.

Corona of 1868.—Bullock.

duced in Mr. Proctor's "Sun," second edition). It shows an extensive system of rifts and rays, which escaped the notice of most observers—their visibility,

THE CORONA. 221

perhaps, depending on the state of the atmosphere, which is described as slightly hazy, but very steady, at Mr. Gilman's station.

The drawings of Captain Tupman and Mr. Foenander (Figs. 71, 72) are interesting for comparison

FIG. 70.

CORONA OF 1869.—SCHOTT.

with each other and with the photographs of the same eclipse (Fig. 73); and that of the eclipse of 1878 (Fig. 74) is remarkable on account of the enormous extension of the faint brushes of nebulosity, which were

222　　　　　　　　　THE SUN.

traced to a distance of 6° or 7° from the sun, by Professors Langley, Abbe, and Newcomb.

One of the first questions which suggests itself with reference to the corona relates to its location: is it a

Fig. 71.

Corona of 1871.—Captain Tupman.

phenomenon of the sun, of the moon, or of our own atmosphere; or is it perhaps a mere optical effect, like a rainbow or a halo? If its seat is in the earth's atmosphere, it is of course an affair of little magnitude or importance; if, on the other hand, it is really at the

THE CORONA. 223

sun, it must be an object of enormous dimensions and of cosmical significance.

Kepler, and many astronomers after him, attributed it to the atmosphere of the moon, and this continued, perhaps, to be the most generally accepted explanation until the early part of the present century, when it was shown by many incontestable considerations that the moon possessed no atmosphere to speak of; certainly

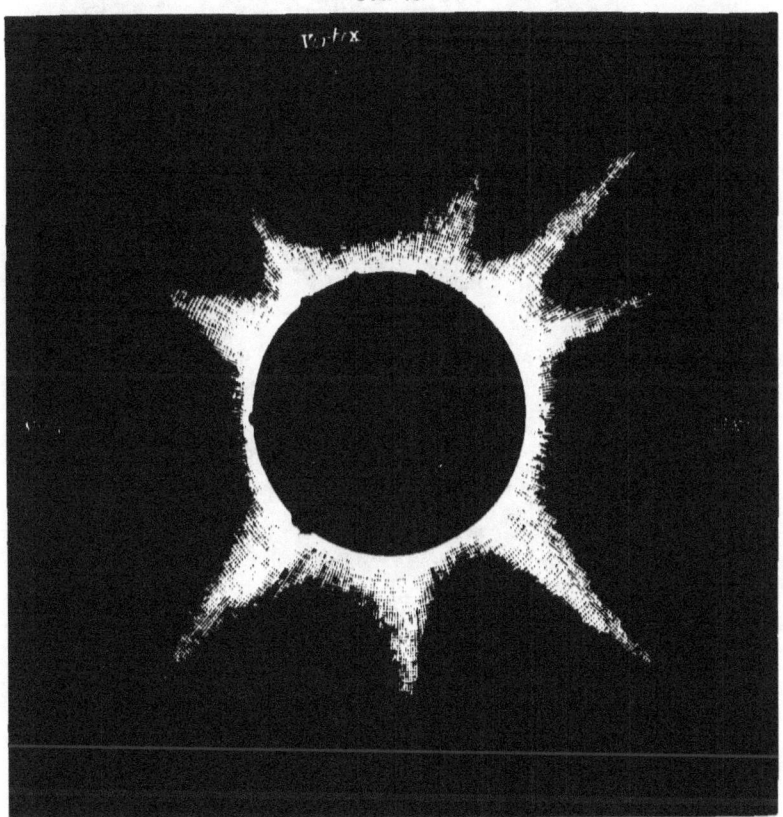

Fig. 72.

Corona of 1871.—Foenander.

none which could account for the observed facts. From this time until 1869 the weight of opinion seems to have been rather in favor of a terrestrial or purely optical origin for the corona, though some (Professor

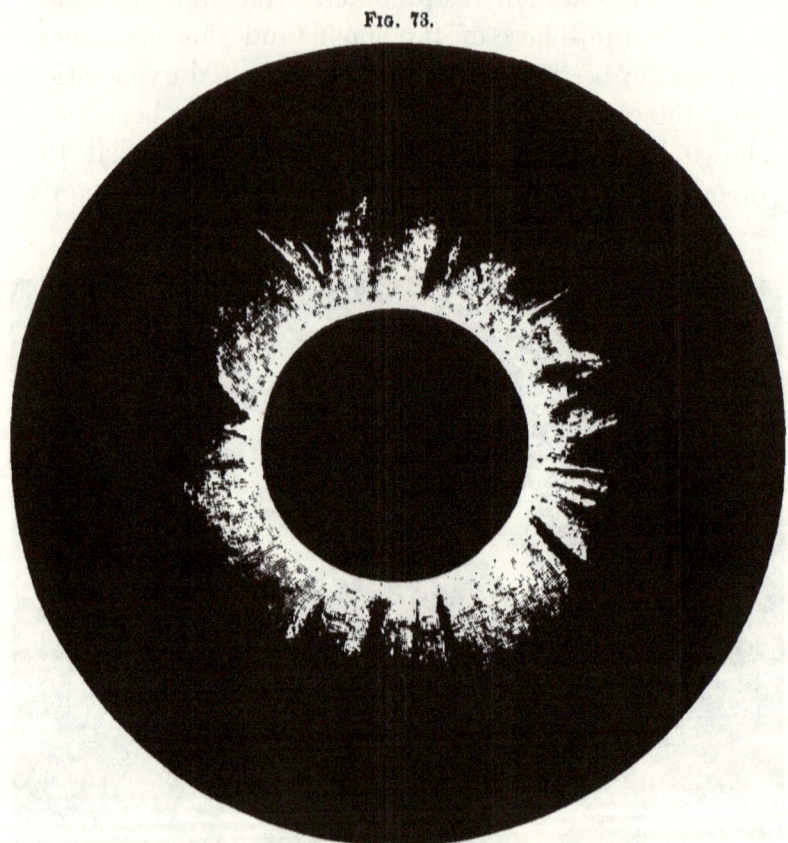

Fig. 73.

Corona of 1871.—From Photographs of Mr. Davis.

Grant, among others, in 1852, in his "History of Physical Astronomy") considered it more probable that the solar atmosphere is the real cause.

The question was first settled in 1869 by the observations of Professor Harkness and the writer, who, independently, found the spectrum of the corona to be characterized by a bright line in the green—the "1,474 line"—so called because on Kirchhoff's map of the solar

THE CORONA. 225

spectrum, then generally used for reference, the line in question falls at this point of the scale. The existence of this bright line demonstrates the presence, in the corona, of incandescent gas, and this of course can only be near the sun. Some doubt was cast upon the observations at first, but they were fully confirmed in 1870; and in 1871 a different and more simple proof was added. Photographs, taken at stations which were separated by several hundred miles, in India and Ceylon, showed precisely the same details of coronal form

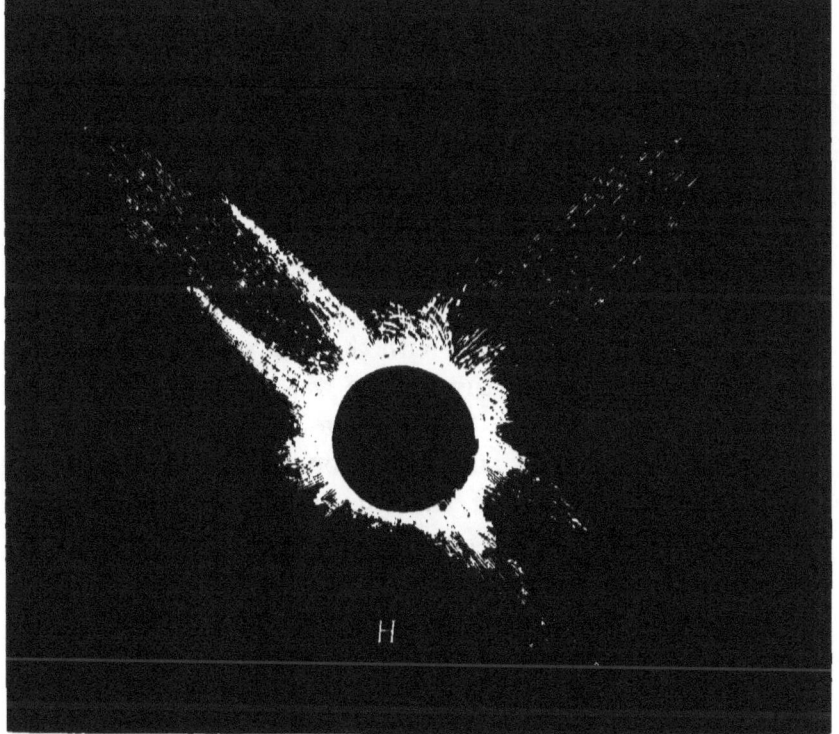

Fig. 74.

CORONA OF 1878.—FROM COMBINATION OF VARIOUS DRAWINGS.

and structure, and are, by themselves considered, sufficient to demonstrate that the main features of the phenomenon are independent of our terrestrial atmosphere and the accidents of the lunar surface. Of course, it is not meant to affirm that our own atmosphere has no part in the phenomenon, but its *rôle* is only secondary. As has been pointed out by Mr. Proctor, the observer at the middle of an eclipse is in the center of an enormous shadow, generally from fifty to a hundred miles in diameter. If we grant that the air retains some sensible density and power of light-reflection, even at an altitude of a hundred miles, and assume for the shadow a radius of only twenty miles, no particle of air illuminated by sunlight could, under these circumstances, be found within 11° of the sun's apparent place in the sky. If there were no corona truly solar in its origin, there would therefore be around the moon a circle of intense darkness, 23° at least in diameter: at the edge of this circle a faint illumination would begin, forming a luminous ring, something like a halo, outside of which the sky would be lighted by rays from an only partially hidden sun. Of course, this dark "hole in the sky" would be concentric with the sun and moon only at the moment when the eclipse was central. In the actual state of things, the portion of the sky in the neighborhood of the sun is, of course, illuminated by whatever appendages of the sun remain unhidden by the moon, and it is this faint illumination, derived from the corona and prominences, which gives to the lunar disk its apparent solid rotundity.

We have spoken of this illumination as faint, but generally it is considered to be much stronger than that of the full moon, though there is some difference of opinion on the matter. There is no doubt that in many cases there is abundant light for reading a watch-face, even at the middle of the totality; the writer, in 1869, found no use for a lantern in making notes or in reading a micrometer-head. But some maintain that the principal portion of this light is derived, not from the corona, but from the illuminated air; for, though the observer himself is in darkness, he has in sight all around the horizon a sunlit atmosphere.

Undoubtedly there is a great difference between different eclipses in respect to the obscurity. The brilliance of the lower part of the corona—a narrow ring close to the limb of the sun—is dazzling; but the light falls off very rapidly. In an eclipse of long duration, therefore, when the moon's apparent diameter considerably exceeds the sun's, the brighter portion of the corona will be covered, and the light will be much less than in an eclipse at a time when the difference between the diameters of the sun and moon is only small.

At the eclipse of 1869 an attempt was made to measure the darkness of the totality, as compared with that of night. The obscurity proved to be so much deeper than had been expected, that the ingenious instrument which Professor Eastman had devised for the purpose turned out inadequate to deal with it exactly. The apparatus consisted of a tube about ten inches long and two and a half in diameter. At the bottom of this was painted a small white star of five points, with a black dot in the center, and a black ring around it. The other end of the tube was closed with a so-called "cat's-eye," a square opening, the size of which can be varied at will, by moving two slides with a micrometer-screw, or rack and pinion.

A small tube, attached obliquely to the large one, like a teapot-nose, allowed the observer to look at the star, and the amount of light from the sky was then measured by opening or shutting the slides until the dot and ring in the center of the star just ceased to be visible. Not only did the ring and dot become invisible with the whole aperture of the cat's-eye, but the star itself was invisible during the totality. Professor Eastman, on the whole, concluded that the general darkness was on this occasion about the same as an hour or so after sunset, when third-magnitude stars first become visible. The instrument was pointed at the zenith, however, and not at the corona, so that it gave no direct determination of the coronal light. Neither do the observations of Mr. Ross, in 1870 (by which the general illumination was compared with the light from a candle), answer the purpose any better.

One or two attempts have been made to compare the shadow cast by the corona with that produced by a candle; but the coronal shadow has always been so masked by the general aërial illumination as to defeat the observation. One astronomer only, so far as known to the writer, has made an estimate of the coronal light based on anything like a scientific foundation. Belli, in 1842, found that the corona seemed to him to give as much light as a candle at a distance of 1·8 metre. He was short-sighted, so that an object like a candle appeared to him as a confused patch of light, and it was by taking advantage of this defect in his vision that he was able to effect the comparison, which must, however, have been only very rough. Two weeks later he compared, in the same way, the full moon, at the same altitude, with a similar candle, and thus found that the light of the corona was less than one sixth that of the moon. This comparison, however, is so unsatisfactory in its details that no great weight can be allowed it, and it must, perhaps, be still considered an open question whether the light of the corona is brighter or not than that of the moon.

The lower portions of the coronal ring, close to the sun, are unquestionably of a brilliance too dazzling to be looked at comfortably with a telescope unprovided with a shade-glass; we have on this point the testimony of Biela, Struve, Ranyard, and others. The same thing is evident from the fact that, at a transit of Venus or Mercury under favorable circumstances, the black disk of the planet becomes visible before it reaches the sun. Janssen thus saw Venus in 1874, and Langley, Mercury in 1878. Of course, this implies behind the planet a background of sensible brightness in comparison with the illumination of our atmosphere. It is generally considered that a difference of one sixty-fourth in the brightness of two adjacent portions of a surface is the smallest quantity perceptible by the eye, and, if so, the corona must be more than one sixty-fourth as bright as the aërial illumination at the edge of the sun's disk. At an eclipse, also, the corona is sometimes seen several seconds, or even minutes, before the beginning and after the end of totality. Petit, in 1860, reports seeing it twelve minutes (*sic*) before the disappearance of the sun, and Lockyer, in 1871, continued to see it for three minutes after the sun's reappearance. But, as has been said before, the light falls off very rapidly, and the outer portions of the corona are of the faintest nebulosity. It is greatly to be desired that, at the next eclipse, some careful photometric measurements should be made.

Apart from the difference in the amount of light at different eclipses, due to the variation in the moon's diameter, there is a strong probability that the corona itself changes considerably in brightness and extent from year to year. In 1878 it was the general verdict of the numerous observers, who had also seen the eclipse of 1869, that the corona was much less brilliant than on the former occasion. Still, several observers of deservedly high reputation hold a precisely contrary opinion. The corona of 1878 was unquestionably the more extensive.

Of course, the known facts as to the periodicity of sun-spots, and the sympathy between them and the prominences, make it antecedently probable that a corresponding variation will be found in the corona; and it is quite certain that, in the eclipse of 1878, which occurred at a sun-spot minimum, the spectroscopic peculiarities of the corona were greatly modified. The bright line, which is its principal characteristic, became so faint that many observers missed it altogether.

This bright line, as has been said before, was first recognized as coronal at the eclipse of 1869. It had been seen reversed in the spectrum of the chromosphere a few weeks previously, both by Mr. Lockyer, and, independently, by the writer, who, however, did not know of the earlier observation until some time after the eclipse. In the ordinary solar spectrum it appears as a fine, dark line at 1,474 of Kirchhoff's scale, or 5,315·9 of Ångstrom's—a line in no way conspicuous as compared with hundreds of others, and barely visible with a single-prism spectroscope. With a spectroscope of high dispersion it was found, in 1876, to be closely double, the upper (more refrangible) component being slightly hazy, while the other is sharp and well-defined. The upper component is the true coronal line, and is always seen without much difficulty, reversed in the spectrum of the chromosphere. Both Kirchhoff and Ångstrom give the line as belonging to the spectrum of *iron*, a fact which was for a time very perplexing, since it is hardly possible that the vapor of this metal could really be the prevailing constituent of the corona, surmounting even hydrogen itself. This difficulty, however, no longer exists, for it is now clear that the iron-line is the lower component of the double, its close proximity to the other being only accidental. The

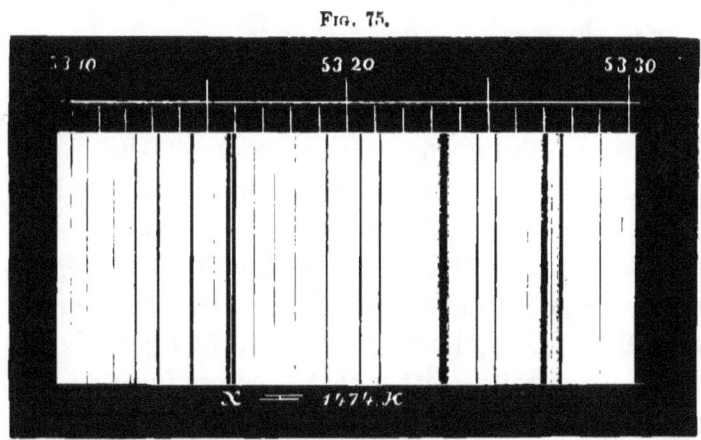

Fig. 75.

Portion of the Spectrum near the Corona Line (*r*), as seen with an instrument of high dispersion.

figure gives a representation of the line and its surroundings, as seen in a high-dispersion spectroscope. The scale above the spectrum is that of Ångstrom.

The hydrogen-lines also appear faintly bright in the spectrum of the corona. It is, perhaps, not quite certain that this may not be due to reflection of the light of the chromosphere in our own atmosphere, but, on the whole, probably not. The atmospheric reflection extends inward, at an eclipse, over the dark disk of the moon, as well as outward, and if the appearance of the hydrogen-lines were due simply to this reflection, they should be just as strong on the moon's disk as in the corona. This does not seem to be the case, though in

1870 the writer saw them plainly on the center of the lunar disk; but Janssen and Lockyer agree that they are much brighter outside. The " 1,474 line " has been traced, by an analyzing spectroscope, on some occasions to an elevation of nearly 20' above the moon's limb, and the hydrogen-lines nearly as far. What is important also, the lines were just as strong *in the middle of a dark rift* as anywhere else. We shall have occasion to recur to this again.

With the analyzing spectroscope the 1,474 line is very much feebler near the sun's limb than the hydrogen-lines, i. e., taking any small portion of the corona near the limb, the hydrogen is much more brilliant than the unknown vapor which produces the other line. When, however, the eclipse is examined by an integrating spectroscope,* the relation of brightness is reversed, showing that the total amount of " 1,474 light " is the greater, and indicating either that it comes from a much more extensive area, or else that in the upper regions the hydrogen loses its brightness much more rapidly than the other material.

As to the substance which produces the 1,474 line we have no knowledge as yet. It would seem that it must be something with a vapor-density far below that of hydrogen itself, which is incomparably the lightest of all bodies known to our terrestrial chemistry. It can hardly be any one of our familiar elements, even in any allotropic modification, such as has been suggested by some, for, in the midst of the most violent disturbances which are observed sometimes in prominences and near sun-spots, when the lines of hydrogen, magnesium, and other metals, are contorted and shattered by the swiftness of the rush of the contending elements, this line

---

* See pages 76, 77 for explanation of this term.

alone remains imperturbable, fine, sharp, and straight; a little brightened, but not otherwise affected. For the present it stands with the line in the extreme red, the so-called helium-line near D, and a few others, as an unexplained mystery.*

Besides this line and the hydrogen-lines, two others have been doubtfully reported in the greenish-yellow part of the spectrum. One of them seems to have been seen twice: first, in 1869 by the writer, and in 1870 by Denza, in Italy. Its place is about 5,570 of Ångstrom's scale. Still, as one of the barium-lines, which is frequently and brilliantly reversed in the spectrum of the chromosphere, is not very far from this place (at 5,534), it is quite possible that this was the line seen. The other doubtful line (reported by the writer in 1869) was at 5,450 (Ångstrom), also very near, in fact between, the places of two lines which are conspicuous in the chromosphere. It will be well to examine the matter more thoroughly at the next opportunity.

Besides bright lines, the corona shows also a faint

---

* Its frequent identification with a line in the spectrum of the aurora borealis, for which, unfortunately, the writer was at first mainly responsible, is a striking example of the difficulty of correcting a mistake which has once gained currency. A few weeks before the first discovery of this line in the spectrum of the corona, Professor Winlock had observed the spectrum of a bright aurora, and had published the position of five lines: one of the five positions coincides with that of the 1,474 line far within the limits of error probable in such an observation, and I jumped to the conclusion that the coincidence was exact and significant. Later observations soon showed that this "line" in the aurora spectrum is not a *line* at all, strictly speaking, but a faint, hazy band, never to be seen except in unusually bright auroras, and not at all identifiable with the 1,474 line of the corona. So far as the spectroscope goes, there is no indication of any connection between the corona and the aurora of the earth's atmosphere, though there are other facts which suggest that the phenomena may be to some extent similar in their nature.

continuous spectrum, and in this Janssen and Barker have observed a few of the more prominent *dark* lines of the solar spectrum—D, *b*, and G especially.

This fact of course shows that while the corona may be in great part composed of glowing gas, as indicated by the bright lines of its spectrum, it also contains a considerable quantity of matter in such a state as to reflect the sunlight—matter, probably, in the form of dust or fog.

This conclusion is borne out also by the result of observations with different forms of polariscope, which, for the most part, indicate that the light of the corona is partially polarized in radial planes, just as it should be if in part composed of reflected light. We have said " for the most part," because there have been some very puzzling discrepancies between different instruments and different observers, which we have not space to discuss here.

Since the corona, then, contains both incandescent gas and also matter in such a condition of mist or smoke as fits it to reflect light, it is an interesting question whether different parts of the coronal structure are composed alike of both, or whether there is a separation.

It has been attempted to solve the question by examining the eclipse with a so-called " slitless spectroscope "—i. e., simply a prism put in front of the objectglass of a small telescope. If, with such an instrument, one were to look at a distant object emitting homogeneous light (an alcohol-flame tinged with salt, for instance), one would see it precisely as if the prism were not there, except that the refraction would change the apparent direction of the object. If the light were composed of three or four bright lines, like that from a

Geissler tube filled with hydrogen, there would then appear the same number of colored images. If the light were like that of an ordinary candle, which gives a continuous spectrum, one would get merely a colored streak. Finally, if we had a source of light combining these different conditions, a lamp-flame, for instance, tinged in some parts with sodium and in others with lithium, we should then have the streak of color marked in the yellow with a clear image of the sodium part of the flame, and in the red and violet with images of that part of the flame which was colored by lithium.

If, then, the long rays and streamers of the corona were mainly composed of the gas which gives the 1,474 line, we ought to see them distinctly through the prism on a background produced by the light from the reflecting mist. Nothing of the kind occurs, however. The slitless spectroscope, in the hands of Respighi and Lockyer in 1871, showed a continuous band of light with several smooth, bright rings upon it: the brightest and largest ring was green (corresponding to the 1,474 line), and there were three other fainter ones in the red, blue, and violet, corresponding to the three brightest lines of hydrogen. It is to be inferred, therefore, that the gaseous matter of the corona forms a pretty regular atmosphere around the sun, and that the structural elements, the rays, rifts, and streamers, are mainly due to mist or dust—at least they seem to give a continuous spectrum. With this agrees the fact, before mentioned, that the 1,474 line is just as bright in the middle of one of the dark rifts as in a bright streamer. In 1878 the slitless spectroscope, however, failed, in the hands of all the observers, to show any rings at all. This fact, taken with the lessened brightness of the corona on that occasion, seems to indicate that the gases of the coronal atmosphere, at the time of a sun-spot minimum, are much diminished in extent and brilliance, while the streamers are comparatively unaffected. It would be easy to speculate upon the significance of this, but one observed instance hardly furnishes a secure enough basis for a sound induction.

The question has been often raised, whether the appearance of the corona changes during an eclipse. Many drawings seem to show that this is the case; they represent the corona at the beginning and end of the eclipse as much wider on that side of the sun less deeply covered by the moon—on the western edge, near the beginning of the eclipse, and on the eastern, near its end—while it is approximately symmetrical at the middle of totality; and this circumstance was much relied upon for a time by those who maintained that the corona is, in the main, a phenomenon of the earth's atmosphere. Other drawings, however, of the same eclipses, show nothing of the kind, nor do the photographs, except in one or two instances, where a sufficient explanation is to be found in drifting clouds. On the other hand, photographs taken at different moments during an eclipse, and at stations many hundred miles apart, agree so closely as to make it evident that the main features of the corona change only gradually, persisting, as a rule, for hours at least, and perhaps for days and weeks for aught we know. At the same time they do sometimes change *perceptibly*, even in the course of twenty minutes, while the shadow is traveling between stations only a few hundred miles apart. Some have thought they saw rapid movements in the streamers, and have described them as waving and flickering; one or two have even imagined that the corona " whirled like a catherine-wheel." Probably this is mere imagination, though the unsteadiness of the air might give a person unused to astronomical observation the idea of scintillating motion. The usual impression upon the mind is quite different—that of calm, serene stability.

Combining the facts that have been ascertained, and speaking in the most general way, it would seem that the corona is mainly composed of filaments which either emanate from the sun or are developed in his atmosphere most abundantly at those portions of his surface about midway between the equator and the poles, those filaments which are emitted on either side of the zone having a tendency to lean toward the central ones. As a consequence, the corona tends toward the form of a four-rayed star, the points of which are inclined 45° to the sun's axis, and are made up of converging filaments, constituting the synclinal structure which Mr. Ranyard first clearly brought out.

Obviously, however, this statement must be taken very loosely. Every eclipse presents striking exceptions. There are always streamers tangential, curved, or inclined, which can be brought under no such rule; faint, far-reaching cones of light, like those which were seen in 1878; dark rifts, rounded masses of nebulosity, vortices, and a multitude of other peculiarities of structure no more reducible to a formula than the shapes of flame or cloud.

Opinion is very widely divided as to the nature and origin of the substances which compose the coronal structures. Every one now, we think, admits the presence of an atmosphere of incandescent gases reaching to an elevation of at least 300,000 miles, and this although there are enormous difficulties in harmonizing an atmosphere of such extent with the low pressure at the surface of the photosphere, indicated by the fineness of the Fraunhofer lines in the spectrum. But, as to the material of which the streamers are composed, and the nature of the forces which determine their form and position, there is no agreement. Some see in the corona simply flocks of meteors, and there can be no doubt that meteoric matter must abound in the sun's immediate neighborhood. But looking, for instance, at the picture of the eclipse of 1871 it appears evident that the details of that corona could not be accounted for in this way. It seems much more likely that the phenomena of comets' tails and the streamers of the aurora are phenomena of the same order, and though as yet the establishment of this relation would not amount to anything like an explanation of the corona, it would be a step toward it—a step by no means taken yet, however, it must be admitted; nor is it easy to see at present how the problem is to be attacked. That the forces concerned reside in the sun himself is made probable by the usual approximate symmetry of the corona with reference to his axis, and the fact that the coronal streamers seem to originate most abundantly nearly in the sun-spot zones.

But we must evidently wait a while for the solution of the problems presented by the beautiful phenomenon. Possibly the time may come when some new contrivance may enable us to see and study the corona in ordinary daylight, as we now do the prominences. The spectroscope, indeed, will not accomplish the purpose, since the rays and streamers of the corona give a continuous spectrum; but it would be rash to say that no means will ever be found for bringing out the structures around the sun which are hidden by the glare of our atmosphere. Unless something like this can be done, the progress of our knowledge must probably be very slow, for the corona is visible only about eight days in a century, in the aggregate, and then only over narrow stripes on the earth's surface, and but from one to five minutes at a time by any one observer.*

With such limited opportunities of observation, it is hardly possible that we should penetrate the mystery very rapidly.

---

* This estimate is based upon the fact that total eclipses occur on the average about once in two years, that the shadow occupies (on the average, again) some three hours in traversing the globe, and that the mean duration of totality is between two and three minutes, never by any possibility reaching eight minutes, and very seldom six.

CHAPTER VIII.

THE SUN'S LIGHT AND HEAT.

Sunlight expressed in Candle-Power.—Method of Measurement.—Brightness of the Sun's Surface.—Langley's Experiment.—Diminution of Brightness at Edge of the Sun's Disk.—Hastings's View as to Nature of the Absorbing Envelope.—Total Amount of Absorption by Sun's Atmosphere.—Thermal, Luminous, and Actinic Rays: their Fundamental Identity and Differences.—Measurement of the Sun's Radiation.—Herschel's Method.—Expressions for the Amount of Sun's Heat.—Pouillet's Pyrheliometer.—Crova's.—Violle's Actinometer.—Absorption of Heat by Earth's Atmosphere; by the Sun's.—Question as to Differences of Temperature on Different Portions of Sun's Disk.—Question as to Variation of Sun's Radiation with Sun-Spot Period.—The Sun's Temperature—Actual—Effective.—Views of Secchi, Ericsson, Pouillet, Vicaire, and Rosetti.—Evidence from the Burning-Glass.—Langley's Experiment with the Bessemer "Converter."—Permanency of Solar Heat for last Two Thousand Years.—Meteoric Theory of Sun's Heat.—Helmholtz's Contraction Theory.—Possible Past and Future Duration of the Sun's Supply of Heat.

SUNLIGHT is the intensest radiance at present known. It far exceeds the brightness of the calcium-light, and is not rivaled even by the most powerful electric arc. Either of these lights interposed between the eye and the surface of the sun appears as a black spot upon the disk.

We can measure with some accuracy the total quantity of sunlight, and state the amount in "candle-power"; the figure which expresses the result is, however, so enormous that it fails to convey much of an idea to the mind—it is 6,300,000,000,000,000,000,000,000 :—six thousand three hundred billions of billions, enumerated in the English manner, which requires a million million to make a billion; or six octillion three hundred septillion, if we prefer the French enumeration.

The "candle-power," which is the unit of light generally employed in photometry,* is the amount of light given by a sperm-candle weighing one sixth of a pound, and burning a hundred and twenty grains an hour. An ordinary gas-burner, consuming five feet of gas hourly, gives, if the gas is of standard quality, from twelve to sixteen times as much light. The total light of the sun is therefore about equivalent to four hundred billion billion of such gas-jets.

This statement rests mainly upon the measurements made by Bouguer in 1725, and Wollaston in 1799; since then, however, confirmed by others. They found that the sun in the zenith would illuminate a white surface about sixty thousand times as intensely as a standard candle at the distance of one metre. Allowing for absorption of light in our atmosphere, the figure would rise to about seventy thousand. As the distance of the sun is ninety-three million miles, or very nearly a hundred and fifty million kilometres, it follows that, if we multiply 70,000 by the square of 150,000,000,000 (reducing kilometres to metres), the product will express the number of candles which, placed on a plane surface facing the earth at the sun's distance, would give a light equal to that of the sun. The number comes out fifteen hundred and seventy-five billion billions (English). It is only necessary further to remember that the surface of a flat disk, such as the sun appears to be, is one fourth of the total surface of a sphere of the same

---

* The French employ a unit just ten times as large—the "Carcel burner."

242   THE SUN.

diameter. We must therefore multiply the above number by four, to obtain that which was stated as the measure of the sun's total light. The number is undoubtedly uncertain by a considerable percentage. It depends upon old observations, which ought to be repeated; observations, also, which are difficult and never very satisfactory because of the vagueness of the unit, the extreme difference between the intensity of the lights compared, and, what is still more troublesome, the difference between the color of the sunlight and of candle-light.

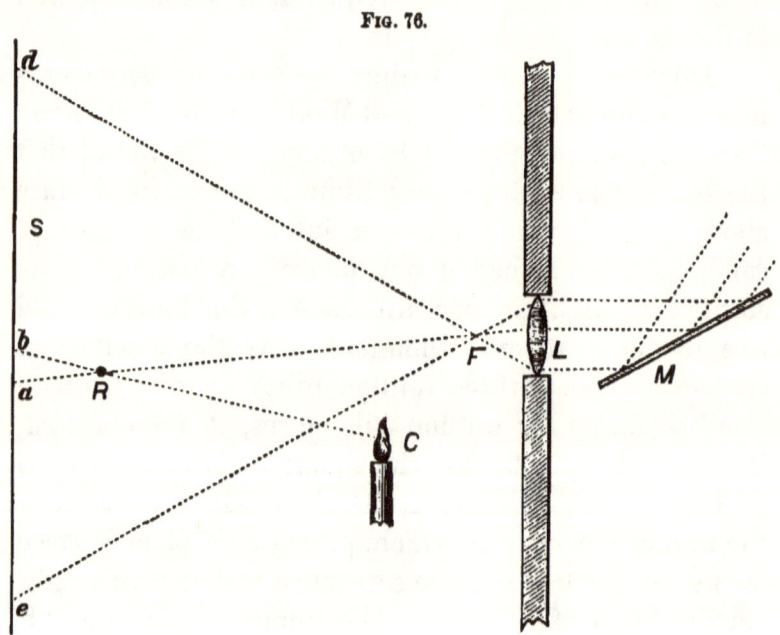

FIG. 76.

METHOD OF MEASURING THE INTENSITY OF SUNLIGHT.

The method of making such a comparison is illustrated by Fig. 76. A mirror, *M*, throws the rays of the sun into a darkened room upon a small lens, the diameter of which is accurately known. This lens brings the rays to a focus at F, after passing which point they diverge and fall upon a white screen, S, at a considerable distance. Neglecting for the present the loss of light by reflection at the surface of the mirror and transmission through the lens, we may say that the illumination of the screen is as many times less than that of full sunlight as the area of the lens L is less than that of the whole disk of light upon the screen. If, for instance, the lens is one fourth of an inch in diameter, and the circle of light on the screen is ten feet across, then the light on the screen would be 23,040 times fainter than sunlight. If we allow for the loss by reflection and in the lens, the ratio would probably not be far from 30,000 to 1. Of course, these two corrections must be (and can be) accurately determined by special observations for the purpose. Having got thus far, there are various methods of proceeding. The simplest, and by no means the least accurate, is to place a small rod, like a pencil, near the screen, so that its shadow will be cast by the sunlight at a: the candle of comparison, C, is then moved back and forth until a position is found at which the shadow cast by its flame at b is equally strong with the other shadow. Then the relative amounts of illumination on the screen produced by the sun and by the candle will be as the squares of the lines $a\,F$ and $b\,C$. There are other methods admitting of still greater precision, but all embarrassed (as this is) by the difference of color between sun- and candle-light. The most uncertain part of the operation lies, however, in the corrections for loss of light in the atmosphere, at the mirror, and in the lens.

Thus far we have considered only the total light emitted by the sun. The question of the intrinsic brightness of his surface is a different though connected one, depending for its solution upon the same observations, combined with a determination of the light-radiating areas in the different cases. Since a candle-flame at the distance of one metre looks considerably larger than the disk of the sun, it is evident that it must be a good deal more than seventy thousand times less brilliant. In fact, it would have to be at a distance of about 1·65 metres to cover the same area of the sky as the sun does, and therefore the solar surface must exceed by a hundred and ninety thousand times the average brightness of the candle-flame.

In the calcium-light the luminous point is both much more brilliant and much smaller than a candle-flame, so that the discrepancy is considerably less. According to certain experiments by Foucault and Fizeau in 1844, the solar surface was found to be a hundred and forty-six times more brilliant than the incandescent lime. At the same time they experimented upon the electric arc, and found the brightest part of this to be only about four times fainter than the sun. Their experiments were, however, conducted by exposure of a Daguerreotype-plate to the rays to be compared, and there is room for considerable doubt as to their accuracy. Later experiments have showed in some cases a rather higher intensity for the brightness of the positive carbon of the electric arc (which is always much more brilliant than the negative). It is asserted in a few instances to have reached a brilliance fully half as great as that of the solar surface; but the evidence is not entirely satisfactory, the comparisons being only indirect. The magnificent lights produced by the dynamo-electric machines of the present day differ from that employed by Foucault and Fizeau, not so much in *intensity* as in *quantity*. The illuminating surfaces are larger, and the extent of the arc much greater, but the brightness of the luminous points concerned seems to remain pretty much the same, and probably depends mainly upon the physical characteristics of the carbon, which are essentially the same in all cases.

One of the most interesting observations upon the brightness of the sun is that of Professor Langley, who, a few years ago (in 1878), made a careful comparison between the solar radiation and that from the blinding surface of the molten metal in a Bessemer "converter." The brilliance of this metal is so great that the dazzling stream of melted iron, which, at one stage of the proceedings, is poured in to mix with the metal already in the crucible, "is deep brown by comparison, presenting a contrast like that of dark coffee poured into a white cup." The comparison was so conducted that, intentionally, every advantage was given to the metal in comparison with the sunlight, no allowances being made for the losses encountered by the latter during its passage through the smoky air of Pittsburg to the reflector which threw its rays into the photometric apparatus. And yet, in spite of all this disadvantage, the sunlight came out *five thousand three hundred* times brighter than the dazzling radiance of the incandescent metal.

Thus far we have spoken of the sun as a whole, but, as has been said before, there is a marked diminution of the light at the edges of the disk; so marked, indeed, that it is exceedingly surprising that any person should ever have questioned the fact, as some—Lambert, for instance—have done. Arago came very near it, for he set the difference at only $\frac{1}{41}$—so little as to be hardly perceptible. An image of the sun a foot in diameter, formed by a small telescope of two inches' aperture, upon a white paper screen, shows the fact, however, in an entirely unquestionable manner. Many measurements have been made for the purpose of comparing the brightness of different parts of the disk. Professors Pickering and Langley, in this country, and Vogel, in Germany, are among the most recent and reliable investigators of the subject. Professor Pickering effected his measurements by forming, with a small telescope, an image of the sun, about sixteen inches across, upon a white screen perforated with an orifice three fourths of an inch in diameter. The telescope was placed horizontally, and the light directed upon it by a mirror, much as in the preceding figure, except that the mirror was moved by clock-work, so as to keep the image constantly in one place. After the rays passed the orifice in the screen they were received upon the disk of a Bunsen photometer, and the light compared with that of a standard candle, in the ordinary way, and thus the ratio was found between the brilliance of the center of the disk and that of other parts. Pickering makes the ratio between the intensity of the light from the edge and center to be thirty-seven per cent.

Vogel, in 1877, proceeded still more elaborately. His instrument, called a spectral photometer, enabled him to compare with great accuracy, and directly, the brightness of the rays of different colors proceeding from different parts of the sun—the red rays by themselves, and the same with the yellow, green, blue, and violet. The following table contains an abridgment of his results. In the first column, headed D, is given the distance of the point from the sun's center in percentage of the sun's radius. The other columns give the ratio between the light of the given color at the center of the disk and at the point in question, expressed also as a percentage. Thus, at the very edge of the disk, at a distance of one hundred per cent. of the sun's radius from its center, the violet light has an intensity of only thirteen per cent. of its intensity at the center, and the red thirty per cent. of *its* central intensity:

D.	Violet. λ 408.	Blue. λ 470.	Green. λ 512.	Yellow. λ 589.	Red. 6l2.	Pickering, general light.
0	100	100	100	100	100	100
10	99·6	99·7	99·7	99·8	99·9	98·8
20	98·5	98·8	98·7	99·2	99·5	....
30	96·3	97·2	96·9	98·2	98·9	....
40	93·4	94·1	94·3	96·7	98·0	94·0
50	88·7	91·3	90·7	94·5	96·7	91·3
60	82·4	87·0	86·2	90·9	94·8	87·0
70	74·4	80·8	80·0	84·5	91·0	....
75	69·4	76·7	75·9	80·1	88·1	78·8
80	63·7	71·7	70·9	74·6	84·3	....
85	56·7	65·5	64·7	67·7	79·0	69·2
90	47·7	57·6	56·6	59·0	71·0	....
95	34·7	45·6	44·0	46·0	58·0	55·4
100	13·0	16·0	18·0	25·0	30·0	37·4

We have added, in a last column, some of the results of Professor Pickering, which, it will be seen, for the most part are in quite satisfactory accordance with those of Vogel.

One thing is obvious from Vogel's table, namely, that the color of the light must be different at the edge of the disk from what it is in the center, since more of the violet light than of the red is lost at the limb.

Professor Langley, in 1875, in attempting to measure directly the relative brightness of points near the center and limb by bringing, in a very ingenious manner, the light from the two points to confront each other on a Bunsen photometer-disk, found this to be a very noticeable fact—the edge is of a sort of chocolate-brown and the center quite bluish, if we take ordinary sunlight as the standard of whiteness. The difference of tint was sufficiently decided to make the measures very difficult. We have never seen in print the results of this work of his, and do not know whether they have yet been published. Vogel's work, however, from the greater completeness of its analysis in respect to the different colors, must take the precedence of everything hitherto done in this line.

The cause of this enfeeblement of the light near the limb of the sun is, of course, the absorption of a portion of the rays by the solar atmosphere.* It becomes, therefore, an interesting subject of inquiry, how much of the sunlight is thus absorbed—how much brighter the sun would shine if suddenly stripped of its gaseous envelopes?

Unfortunately, the question does not, in the present state of science, admit of a certain and definite answer. By making certain assumptions as to the constitution of the luminous surface and the character of the atmosphere we may, it is true, deduce mathematical formulæ

---

* It has generally been considered, hitherto, that this absorbing envelope must be gaseous, and it has usually been identified with the so-called reversing layer which, at an eclipse, gives the bright-line spectrum seen at the beginning and end of totality. Professor Hastings, of Baltimore, has, however, lately proposed a somewhat different theory, viz., that the absorption is produced by something like smoke—i. e., by matter in a pulverulent condition, at a lower temperature than the photospheric clouds, and disseminated through the lower portions of the sun's true atmosphere. He urges with force that the absorption of the gases themselves, at such a temperature, must be *selective*, producing bands and lines in the spectrum, while the absorption with which we have to do in this case is *general*, simply weakening all the rays pretty much alike, though of course affecting those of short-wave length more than those of long, as previously pointed out by Langley. The substance concerned, he says, must be one which condenses and precipitates at a temperature higher than that of the photosphere, so that its *vapor* would not be present to any appreciable extent in the photosphere and reversing layer, and its lines would not be found in the solar spectrum. He suggests that the substance is very probably carbon.

(of a rather complicated character) which will represent the observed facts *on those assumptions.*

Laplace, for instance, assumed that each point upon the luminous surface of the sun radiated equally in all directions, and that its atmosphere was homogeneous throughout—knowing, of course, that it could not be homogeneous, but not knowing what laws of density and temperature would apply in the case, and therefore not being able to supply a more correct hypothesis. On these assumptions, and taking as a basis of calculation the observations of Bouguer, which in the main agree with the more modern ones, he found that the solar atmosphere must absorb about eleven twelfths of the whole light; in other words, that the sun, without its atmosphere, would be about *twelve* times as bright as we see it now. Secchi has also adopted his conclusion.

His first assumption, however, is probably very far from true. So far as we know, no luminous surface behaves as he supposes, but generally the radiations at an oblique angle are vastly less powerful than those perpendicular to the surface. According to Laplace's assumption, the sun, without its atmosphere, would be much brighter at the edge than at the center. Now, an incandescent sphere of metal, or an illuminated globe of white glass (like the shade of a student-lamp), appears sensibly of equal brightness all over, the foreshortening of each square inch of surface inclined to the line of sight just compensating for its diminished radiation. Assuming *this* law of radiation for the solar surface, and still keeping the hypothesis of a homogeneous atmosphere, Professor Pickering shows that the observed darkening from the center to the edge of the sun's disk, indicated by his measures, would be accounted for pretty accurately by supposing this atmosphere to have a height approximately equal to the sun's radius, and of such absorbent power as to reduce the light by about seventy-four per cent. at the center of the disk, leaving twenty-six per cent. to pass. From this it is possible to show that the whole light, if there were no solar atmosphere, would be about four and two thirds times as great as now—always, be it remembered, accepting the assumptions.

Vogel, assuming the same fundamental law of radiation, finds from his observations that the removal of the solar atmosphere would increase the brightness of its red rays about 1·49 times, and of the violet 3·01. The difference between this result and that of Pickering is larger than would be expected from the general near accordance of the observations, but is probably principally due to the fact that Vogel employs a formula of Laplace's which implicitly assumes the solar atmosphere to be very thin as compared with the size of the sun itself, while Pickering's method of calculation accepts no such limitation. There is an important difference also between the observations of the two investigators near the edge of the disk: Vogel's observations show a much more rapid degradation of the light just there, and so indicate a much thinner and denser atmosphere than Pickering's.

It is evident, however, that for the present we must content ourselves with the rather vague statement that the removal of the sun's atmosphere would multiply its brightness several times. It is almost certain that the amount of light received by the earth would be doubled; it is hardly likely that it would be quintupled. Moreover, its color would be materially changed, and its tint, as pointed out by Langley, would be more bluish than now. The solar atmosphere *reddens* the light transmitted through it, in just the same way that our terrestrial atmosphere does at sunset, but to a less degree.

Thus far we have confined ourselves to those radiations which affect the sense of vision. But these rays do more: if received upon a dark surface they are, as we say, "absorbed," and the absorbing body becomes warmer. Nothing in science is now much more certain than that these luminous radiations consist of pulses of inconceivable (but measurable) frequency, which are communicated through intervening space; pulses which are capable not merely of affecting the visual nerves of sentient beings, but of producing also many other effects, physical, thermal, or chemical, according to the surface which receives them. The human eye, however, is very circumscribed in its range of perception, taking cognizance only of such vibrations as do not exceed or fall short of certain limits of frequency—the slowest oscillations it recognizes being those of the extreme red, which performs about three hundred and ninety millions of millions of vibrations in a second; while the most rapid, those of the extreme violet, are nearly twice as frequent, making seven hundred and seventy millions of millions in the same time. The rays emitted by the sun are not, however, so limited; but the visual vibrations are accompanied by others both many times more slow and more rapid. There has been a prevailing idea for many years, founded upon Brewster's fallacious experiments, that thermal, luminous, and chemical rays are fundamentally different, though coexistent in the sun's beams. This is erroneous. It is true, indeed, that rays whose vibrations are too slow to be seen produce powerful heating effects, and that those which are invisible because they are too rapid have a strong influence in determining certain chemical and physical reactions; but it is also true that the visible rays are capable of producing the same effects to a greater or less degree, and there is some reason for thinking that certain animals can see by rays to which the human retina is insensible. There is absolutely no philosophical basis for distinction between the visible and invisible radiations of the sun, except in the one point of vibration-frequency—their *pitch*, to use the analogy of sound. The expressions thermal, luminous, and chemical rays are apt to be misleading. All the waves of solar radiation are carriers of energy, and when intercepted do work, producing heat, or vision, or chemical action, according to circumstances.

If the amount of solar light is enormous, as compared with terrestrial standards, the same thing is still more true of the solar heat, which admits of somewhat more accurate measurement, since we are no longer dependent on a unit so unsatisfactory as the "candle-power," and can substitute thermometers and balances for the human eye.

It is possible to intercept a beam of sunshine of known dimensions, and make it give up its radiant energy to a weighed mass of water or other substance, to measure accurately the rise of temperature produced in a given time, and from these data to calculate the whole amount of heat given off by the sun in a minute or a day.

Pouillet and Sir John Herschel seem to have been the first fairly to grasp the nature of the problem, and to investigate the subject in a rational manner.

Herschel's experiments were made in 1838 at the Cape of Good Hope, where he was then engaged in his astronomical work. He proceeded in this way: A small tin vessel, containing about half a pint of water,

THE SUN'S LIGHT AND HEAT. 253

carefully weighed, was placed on a light wooden support, touching it at only three points. This was put inside of a considerably larger cylinder, also of tinned iron, this outer cylinder having a double cover with a hole in it, the cover large enough to shade the sides of the vessel, and the hole a little less than three inches in diameter. A delicate thermometer was immersed in the water, with a sort of dasher of mica for the purpose of stirring it and keeping the temperature uniform throughout the mass. The apparatus was so placed and adjusted that the whole of the light and heat passing through the hole in the cover would fall upon the surface of the water, the sun at that time (December 31st) being within 12° of the zenith at noon.

This apparatus was placed in the sunshine and allowed to stand for ten minutes, shaded by an umbrella, and the slight rise in the temperature of the water was noted. Then the umbrella was removed and the solar rays were allowed to fall upon the water for the same length of time, and the much larger rise of temperature was noted. Finally, the apparatus was again shaded, and the change for ten minutes again observed. The mean between the effects in the first and last ten-minute intervals could be taken as the measure of the influence of other causes besides the sun, and deducting this from the rise during the ten minutes' insolation, we have the effect of the simple sunshine.

Herschel's figures for his first experiment run as follows:

Rise of temperature in first ten minutes.............. 0·°25
  "   "   "   " second ten minutes (sun)........ 3·°90
  "   "   "   " third ten minutes .... 0·°10

The mean of the first and third is 0·°17, and this deducted from the second gives 3·°73 as the rise of temperature produced by a sunbeam three inches in diameter, absorbed by a mass of matter equivalent to 4,638 grains of water (we do not indicate the minutiæ of the process by which the weight of the tin vessel, thermometer, stirrer, etc., are allowed for). Nothing more is now necessary to enable us to compute just how much heat is received by the earth in a day or a year, except, indeed, the determination of the very troublesome and somewhat uncertain correction for the absorption of heat by the earth's atmosphere—a correction deduced by means of observations made at varying heights of the sun above the horizon.

Herschel preferred to express his results in terms of melting ice, and put it in this way: the amount of heat received on the earth's surface, with the sun in the zenith, would melt an inch thickness of ice in two hours and thirteen minutes nearly.

Since there is every reason to believe that the sun's radiation is equal in all directions, it follows that, if the sun were surrounded by a great shell of ice, one inch thick and a hundred and eighty-six million miles in diameter, its rays would just melt the whole in the same time. If, now, we suppose this shell to shrink in diameter, retaining, however, the same quantity of ice by increasing its thickness, it would still be melted in the same time. Let the shrinkage continue until the inner surface touches the photosphere, and it would constitute an envelope more than a mile in thickness, through which the solar fire would still thaw out its way in the same two hours and thirteen minutes—at the rate, according to Herschel's determinations, of more than forty feet a minute. Herschel continues that, if this ice were formed into a rod 45·3 miles in diameter, and darted toward the sun with the velocity of light, its advancing point would be melted off as fast as it approached, if by any means the whole of the solar rays could be concentrated on the head. Or, to put it differently, if we could build up a solid column of ice from the earth to the sun, two miles and a quarter in diameter, spanning the inconceivable abyss of ninety-three million miles, and if then the sun should concentrate his power upon it, it would dissolve and melt, not in an hour, nor a minute, but in a single second: one swing of the pendulum, and it would be water; seven more, and it would be dissipated in vapor.

In formulating this last statement we have, however, employed, not Herschel's figures, but those resulting from later observations, which increase the solar radiation about twenty-five per cent., making the thickness of the ice-crust which the sun would melt off of his own surface in a minute to be much nearer fifty feet than forty.

To put it a little more technically, expressing it in terms of the modern scientific units, the sun's radiation amounts to something over a million *calories* per minute for each square metre of his surface, the *calory*, or heat-unit, being the quantity of heat which will raise the temperature of a kilogramme of water one degree centigrade.

An easy calculation shows that, to produce this amount of heat by combustion would require the hourly burning of a layer of anthracite coal more than sixteen feet (five metres) thick over the entire surface of the sun—nine tenths of a ton per hour on each square foot of surface—at least nine times as much as the consumption of the most powerful blast-furnace known to art. It is equivalent to a continuous evolution of about ten thousand horse-power on every square foot of the sun's whole area. As Sir William Thomson has shown, the sun, if it were composed of solid coal, and produced its heat by combustion, would burn out in less than six thousand years.

Of this enormous outflow of heat the earth of course intercepts only a small portion, about $\frac{1}{2200000000}$. But even this minute fraction is enough to melt yearly, at the earth's equator, a layer of ice something over one hundred and ten feet thick. If we choose to express it in terms of "power," we find that this is equivalent, for each square foot of surface, to more than sixty tons raised to the height of a mile; and, taking the whole surface of the earth, the *average* energy received from the sun is over fifty mile-tons yearly, or one horse-power continuously acting, to every thirty square feet of the earth's surface. Most of this, of course, is expended merely in maintaining the earth's temperature; but a small portion, perhaps $\frac{1}{1000}$ of the whole, as estimated by Helmholtz, is stored away by animals and vegetables, and constitutes an abundant revenue of power for the whole human race.*

If we inquire what becomes of that principal portion of the solar heat which misses the planets and passes off into space, no certain answer can be given. Remem-

* Several experimenters have contrived machines for the purpose of utilizing the solar heat as a source of mechanical energy, among whom Ericsson and Mouchot have been most successful. M. Pifre describes, in a recent number of the "Comptes Rendus," some results from a machine of Mouchot's construction, claiming to have utilized more than eighty per cent. of the heat which falls on the mirrors of the instrument—something over twelve calories to a square metre. We do not mean, of course, that this percentage of the total solar energy appeared as mechanical power in the *engine*, but only in its *boiler*. The machine had a mirror-surface of nearly a hundred square feet, and gave not quite a horse-power. It is very possible that such machines will find useful application in the rainless regions like Egypt and Peru.

bering, however, that space is full of isolated particles of matter (which we encounter from time to time as shooting-stars), we can see that nearer or more remotely in its course each solar ray is sure to reach a resting-place. Some have attempted to maintain that the sun sends heat only toward its planets; that the action of radiant heat, like that of gravitation, is only *between* masses. But all scientific investigation so far shows that this is not the case. The energy radiated from a heated globe is found to be alike in all directions, and wholly independent of the bodies which receive it, nor is there the slightest reason to suppose the sun any way different in this respect from every other incandescent mass.

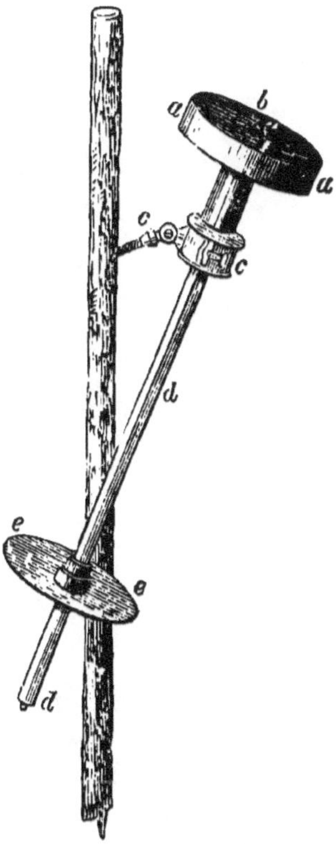

Pouillet's experiments were made about the same time as Herschel's, but with a different apparatus, though based on the same principles. He named his instrument the pyrheliometer, or "measurer of solar fire." Fig. 77 represents it. The little snuffbox-like vessel, a, b, of silver-plated copper, blackened on the upper surface, contains a weighed quantity of water, and a thermometer is immersed in it, the mercury in its stem being visible at d. The disk, e, e, makes it easy to point the instrument squarely to the sun, by directing it so that the shadow of a falls concentrically upon this disk. The button at the lower end

Fig. 77.

is for the purpose of agitating the water in the vessel a, a, by simply turning the whole thing on its axis, in the collar c, c. The instrument is much more convenient than Herschel's apparatus, but hardly as accurate, except under very careful manipulation.

Crova has modified it by filling the upper vessel with mercury, which is a better conductor of heat than

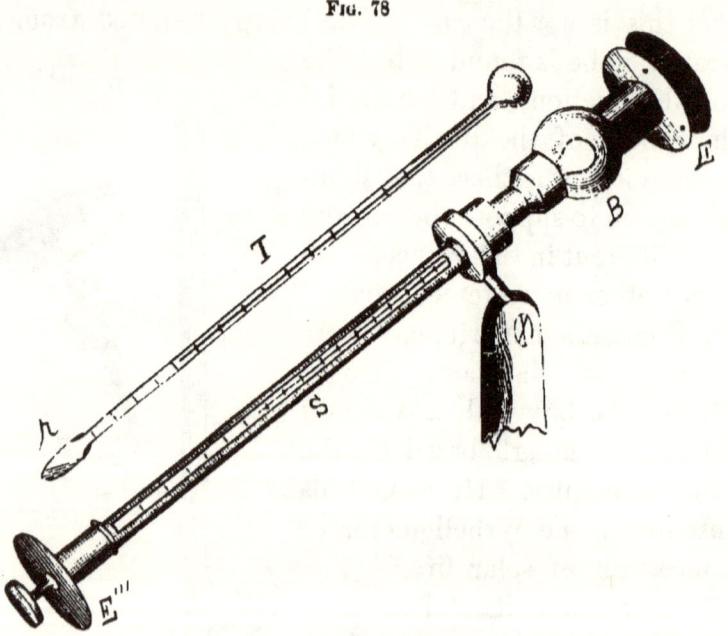

FIG. 78

CROVA'S PYRHELIOMETER.

water. For relative measurements, as, for instance, a comparison of the amounts of heat received from the sun at different hours of the day, Crova employs a slightly different instrument, of which Fig. 78, copied from his paper in the "Annales de Chimie," for February, 1880, is a reproduction.

An exceedingly sensitive alcohol thermometer, shown separately at T, with a large bulb carefully blackened, is inclosed in a double-walled sphere, B, nickel-plated on the outside. An opening in the walls of the sphere, carefully aligned with a similar opening in a double screen, E, allows a beam of light to fall upon the thermometer-bulb, the beam being about two thirds the diameter of the bulb. The thermometer is constructed with a supplementary reservoir, r, at the lower end, by means of which the end of the indicating column can be made to fall near the middle of the scale at any temperature, the object being to measure only *changes* of temperature, not absolute temperatures. The bulb and tube are so proportioned that a degree on the scale is nearly half an inch long, thus permitting great accuracy of reading. In order, however, to determine just how much heat is required to raise the thermometer of this instrument 1°, it is necessary to compare it with one of the standard instruments, by exposing it to the sun at the same time.

This method of procedure, by which we determine the rate at which a sunbeam of given dimensions communicates heat to a measured mass of matter, is known as the *dynamic* method. It is somewhat inconvenient in requiring considerable time and a number of readings.

There is a different process for deducing the same results, which has been employed by Waterston, Ericsson, Secchi, Violle, and others, and may be called the *statical* method. It consists essentially in observing how much the sun will raise the temperature of a body exposed to its rays above that of the inclosure in which it is placed, this inclosure being kept at a fixed and known temperature by the circulation of water, or some such means. Instruments based on this principle are called *actinometers*. Of these, probably the most complete in its arrangements is that of Violle, described in his paper upon the mean temperature of the sun's surface, published in the "Annales de Chimie," in 1877. We give a diagram of the instrument. It consists of two concentric spheres of thin metal, the outer twenty-three centimetres in diameter, the inner fifteen centimetres. The outer is polished on the outside; the

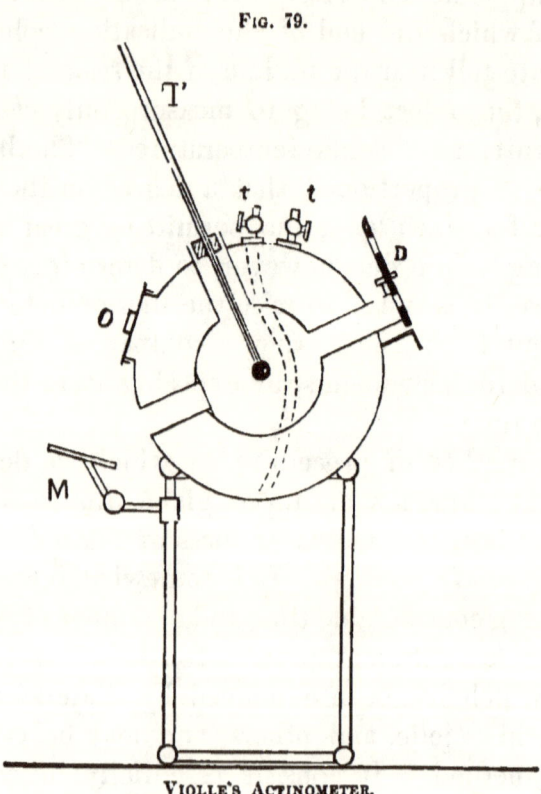

Fig. 79.

VIOLLE'S ACTINOMETER.

inner is blackened on the inside. The space between the two spheres is filled with water, which is kept at a uniform temperature either by mixing snow or ice with it, or else by a current circulated through it by means of the stopcocks t, t. A sensitive thermometer, T, has its blackened bulb placed in the center of the inner sphere, the stem reaching outside through a tubulure provided for the purpose. Two opposite openings, shown in the figure, allow a beam of sunlight to pass through the globes. A perforated screen at D limits its diameter, so that none of it shall touch the walls of the vessel, though the thermometer-bulb is entirely covered by it. A small screen at M allows the observer to see the shadow of the thermometer-bulb, and so to perceive whether the tube through which the light enters is properly directed. If the apparatus is mounted upon what is called an equatorial stand, like a telescope, and provided with clock-work, the whole labor of observation will consist merely in reading the thermometer. The difference between its temperature and that of the water in the surrounding shell gives the necessary data for calculating the intensity of the solar radiation at the time of reading, since the heat received by the thermometer from the sun and shell together must just equal that radiated back by the thermometer-bulb to the shell, after allowing for the orifices.

Violle found that at noon on a fair day the thermometer of this apparatus generally stood, when exposed to the sun, from $10.5°$ to $12.5°$ centigrade (i. e., $18.9°$ to $22.5°$ Fahr.) above the temperature of the shell when the latter was filled with ice-water. If it was filled with boiling water, as in some of his experiments, the difference became less by about $1°$ centigrade.

The results obtained with instruments of this class, of course, agree very closely with those reached by the dynamic method.

It need hardly be said that the amount of heat received from the sun in a minute, by a given area exposed to its radiation, varies widely, according to the altitude of the sun and the condition of the air; indeed, the most difficult part of the experimental problem lies in the determination of the corrections to be applied on account of the absorption of the earth's atmosphere. It would take us too far to discuss the formulæ and methods of calculation which have been proposed. They are necessarily very complicated (those, at any rate, which are tolerably accurate in their results), because they have to take into account the meteorological conditions, especially the hygrometric state of the air. Besides this, the absorption varies greatly for radiations of different pitch, so that the violet rays, which are photographically the most active, suffer more than the green and yellow, which are most effective in the growth of plants; and these more than the red; and the red, in their turn, much more than the low-pitched, slowly vibrating waves which, though invisible, are still powerful carriers of energy.

Speaking loosely, it may be estimated that, at the sea-level, in fair weather, neither excessively moist nor dry, about thirty per cent. of the solar radiation is absorbed when the sun is at the zenith, and at least seventy-five per cent. at the horizon. Of the rays striking the upper surface of the atmosphere, between forty-five and fifty per cent., therefore, are generally intercepted in the air, even when there are no clouds.

Of course, it does not follow that the heat absorbed in our atmosphere is lost to the earth. Far from it: the air itself becomes warmed and communicates its heat to the earth; and, since the atmosphere intercepts a large proportion of the heat which the earth would radiate into space if not thus blanketed, the temperature of the earth is kept much higher than it would be if there were no air.

Instead of stating how much ice would be melted in

a minute by a given sunbeam, we may give the number of calories received per minute by one square metre exposed perpendicularly to the sun's rays at the upper surface of the atmosphere. This number, which measures the sun's radiation, is called the "solar constant," and, according to different experimenters, ranges from Pouillet's estimate, 17·6, to that of Forbes, who found 28·2. The most reliable recent determinations by Crova and Violle set it at 23·2 and 25·4 respectively. Probably 25 is very near the truth, since the results obtained by the same observer on different days, after all possible pains is taken with the corrections, are even more discordant than the numbers given above.* Very possibly a continued series of observations at some very elevated station would considerably improve the data.

Experiments with the thermopile show that the heat radiated by the solar disk varies, like the light, very considerably from the center to the edges. The first observations of this kind were made by Professor Henry at Princeton in 1845, and have since been repeated by many others, Secchi and Langley especially. According to Langley, the heat emitted from a point about 20″ from the limb is only one half that from the same extent of surface at the center of the disk. His table runs as follows, the first column giving the distance from the center of the disk, and the second the intensity of radiation shown by the thermopile:

Distance from center.	Heat-radiation.
0·00	100
0·25	99
0·50	95
0·75	86
0·95	62
0·98	50

* The recent investigations of Langley (for which see Appendix) seem to indicate that even the result of Forbes is none too high.

If we compare this table with that given on a preceding page, which gives the variation of luminosity from center to edge of the solar disk, it is at once evident, as Langley was the first to point out, in 1875, that the absorption is, to a certain extent, selective, the short waves of the solar radiation being more affected than the long. Besides this regular variation of the radiation from center to edge, Secchi, in 1852, found, or thought he found, a notable difference between the radiation from the equator of the sun and that from the higher latitudes, the difference being at least one sixteenth between the equator and latitude 30°. The northern hemisphere he also found to be a little hotter than the southern. Later investigators (Langley especially) have failed to find any such difference; and on the whole it seems probable that Secchi was mistaken, though this is not certain, as it would be quite unsafe to assert that the actual condition of the sun's surface may not have changed between 1852 and 1876.

In connection with the absorption of the solar atmosphere, Langley has ventured some interesting speculations. After showing that variations in the number and magnitude of sun spots can not *directly* produce any sensible effect upon terrestrial temperatures, he calls attention to the fact that even slight changes in the depth and density of the sun's absorbing layer would make a great difference; and he raises the question whether we may not find here the explanation of glacial and carboniferous periods in the earth's history. It is quite certain that, were the envelope removed, the solar radiation would be at least doubled, and perhaps increased in a much higher ratio, while any considerable increase of its thickness would so diminish our heat-supply as to give us perpetual winter.

As yet our means of observation have not sufficed to detect with certainty any variations in the amount of heat emitted by the sun at different times. That there are such variations is almost certain, since the nuclei of sun-spots radiate much less heat, as well as light, than neighboring regions of the solar surface, and the faculæ more: this has been directly determined with the thermopile. The whole amount of variation in the total heat-supply has, however, proved too small for measurement with our present instruments, and science waits anxiously for apparatus and methods of delicacy adequate to deal with the problem. As was said in the chapter upon the sun-spots, we are as yet entirely uncertain whether, at the time of a sun-spot maximum, the solar radiation is more or less powerful than ordinarily.

There has been a great deal of pretty vigorous discussion as to the temperature of the sun, and that the subject is a difficult one is evident enough from the wide discrepancy between the estimates of the highest authorities. For instance, Secchi originally contended for a temperature of about 18,000,000° Fahr. (though he afterward lowered his estimate to about 250,000°); Ericsson puts the figure at 4,000,000° or 5,000,000°; Zöllner, Spoerer, and Lane name temperatures ranging from 50,000° to 100,000° Fahr., while Pouillet, Vicaire, and Deville have put it as low as between 3,000° and 10,000° Fahr. The intensest artificial heat may perhaps reach 4,000° Fahr.

The difficulty is twofold. In the first place, the sun can not properly be said to have *a* temperature any more than the earth's atmosphere can. The temperature of different portions of the solar envelope must vary enormously, increasing fast as we descend below the surface, so that in all probability there may be a difference of thousands of degrees between the temperature at the upper surface of the photosphere and that at the sun's center, or even at the depth of a few thousand miles.

We may, however, partially evade this difficulty by substituting as the object of inquiry the sun's *effective* temperature—i. e., instead of seeking to ascertain the actual temperature of different parts of the sun's surface, we may inquire what temperature would have to be given to a uniform surface of standard radiating power (a surface covered with lampblack is generally taken as this standard), and of the same size as the sun, in order that it might emit as much heat as the sun actually does. In this way we obtain a perfectly definite object of investigation. But the problem still remains very difficult, and has obtained as yet no entirely satisfactory solution. The difficulty lies in our ignorance as to the laws which connect the temperature of a surface with the amount of heat radiated per second. So long as the temperature of the radiating body does not greatly exceed that of surrounding space, the heat emitted is very nearly proportional to the excess of temperature. The extremely high values of the solar temperature asserted by Secchi and Ericsson depend upon the assumption of this law (known as Newton's) of proportionality between the heat radiated and the temperature of the radiating mass—a law which direct experiment proves to be untrue as soon as the temperature rises a little. In reality, the amount of heat radiated increases much faster than the temperature.

More than forty years ago the French physicists, Dulong and Petit, by a series of elaborate experiments, deduced an empirical formula, which answered pretty satisfactorily for temperatures up to a dull-red heat. By applying this formula, Pouillet, Vicaire, and others arrived at the low solar temperatures assigned by them. It is, however, evidently unsafe to apply a purely empirical formula to circumstances so far outside the range of the observations upon which it was founded, and, in fact, within a few years several experimenters, Rosetti especially, have shown that it needs modification, even in the investigation of artificial temperatures like that of the electric arc. Rosetti, from his observations, has deduced a different law of radiation, and by its application finds 10,000° Cent., or 18,000° Fahr., as the *effective temperature* of the sun—a result which, all things considered, seems to the writer more reasonable and better founded than any of the earlier estimates. Rosetti considers that this is also pretty nearly the actual temperature of the upper layers of the photosphere. The radiating power of the photospheric clouds, to be sure, can hardly be as great as that of lampblack; but, on the other hand, their radiation is supplemented by that of other layers, both above and below.

Besides the data as to the intensity of the solar temperature obtained by calculation from the measured emission of heat, we have also direct evidence of a very impressive sort. When heat is concentrated by a burning-glass, the temperature at the focus can not rise above that of the source of heat, the effect of the lens being simply to move the object at the focus virtually toward the sun; so that, if we neglect the loss of heat by transmission through the glass, the temperature at the focus should be the same as that of a point placed at such a distance from the sun that the solar disk would seem just as large as the lens itself viewed from its own focus.

The most powerful lens yet constructed thus virtually transports an object at its focus to within about two hundred and fifty thousand miles of the sun's surface, and in this focus the most refractory substances—platinum, fire-clay, the diamond itself—are either instantly melted or dissipated in vapor. There can be no doubt that, if the sun were to come as near us as the moon, the solid earth would melt like wax.

We have spoken, a few pages back, of Professor Langley's experimental comparison between the brilliance of the solar surface and that of the metal in a Bessemer converter. At the same time he made measurements of the heat by means of a thermopile, and found the heat radiation of the solar surface to be *more than* eighty-seven times as intense as that from the surface of the molten metal. It will be recalled that the experiment only sets a lower limit to the solar radiation, so that it is altogether probable that, were all the necessary corrections determined and applied, the ratio would be increased from eighty-seven to at least a hundred, and perhaps to a hundred and fifty. Ericsson, in 1872, made a somewhat similar comparison in a different and exceedingly ingenious manner. He floated a calorimeter containing about ten pounds of water upon the surface of a large mass of molten iron, by means of a raft of fire-brick. The calorimeter was raised a little above the surface, and the water contained was kept in circulation by suitable mechanism. He found that the radiation of the metal was a trifle over two hundred and fifty calories per minute for each square foot of surface. This is equivalent to twenty-seven hundred and ninety calories to the square metre, and is only $\frac{1}{400}$ of the sun's emission. He estimated the temperature of the metal at 3,000° Fahr., or 1,538° Cent. Professor Langley, in his experiment, estimated the temperature of the Bessemer metal much higher—superior, in fact, to the temperature of melting platinum, which is usually considered to be about 2,000° Cent. He bases this conclusion upon the fact that platinum wire, stretched above the mouth of the converter, or dipped into the issuing stream, was immediately melted. Since, however, iron and its vapor attack platinum much in the same way as mercury and its vapor attack gold, there may be some doubt as to the correctness of his estimate. The same conclusions as to the intensity of the solar temperature follow from investigations by Soret and others as to the penetrating power of the sun's rays, and from a comparison with artificial sources of heat in respect to the relative proportion of the rays of different wave-lengths in the total radiation. A body of low temperature emits an enormous proportion of slow-swinging, invisible vibrations, while, as the temperature rises, the shorter waves become proportionally more and more abundant. Thus, in the composition of a body's radiation, we get some clew to its temperature. Hitherto all such tests concur in putting the sun's temperature high above that of any known terrestrial flame.

And now we come to questions like these: How is such a heat maintained? How long has it lasted already? How long will it continue? Are there any signs of either increase or diminution?—questions to which, in the present state of science, only somewhat vague and unsatisfactory replies are possible.

As to progressive changes in the amount of the solar heat it can be said, however, that there is no evidence of anything of the sort since the beginning of authentic records. There have been no such changes in the distribution of plants and animals within the last two thousand years, as must have occurred if there had been, within this period, any appreciable alteration in the heat received from the sun. So far as can be made out, with few and slight exceptions, the vine and olive grow just where they did in classic days, and the same is true of the cereals and the forest-trees. In the remoter past there have been undoubtedly great changes in the earth's temperature, evidenced by geological records—carboniferous epochs, when the temperature was tropical in almost arctic latitudes, and glacial periods, when our now temperate zones were incased in sheets of solid ice, as northern Greenland is at present. Even as to these changes, however, it is not yet certain whether they are to be traced to variations in the amount of heat emitted by the sun, or to changes in the earth herself, or in her orbit. So far as observation goes, we can only say that the outpouring of the solar heat, amazing as it is, appears to have gone on unchanged through all the centuries of human history.

What, then, maintains the fire? It is quite certain, in the first place, that it is not a case of mere combustion. As has been said, only a few pages back, it has been shown that, even if the sun were made of solid coal, burning in pure oxygen, it could only last about six thousand years: it would have been nearly one third consumed since the beginning of the Christian era. Nor can the source of its heat lie simply in the cooling of its incandescent mass. Huge as it is, its temperature must have fallen more than perceptibly within a thousand years if this were the case.

Two different theories have been proposed, which are probably both true to some extent. One of them finds the chief source of the solar heat in the impact of meteoric matter, the other in the slow contraction of the sun. As to the first, it is quite certain that a part of the solar heat is produced in that way; but the question is whether the supply of meteoric matter is sufficient to account for any great proportion of the whole. As to the second, on the other hand, there is no question as to the adequacy of the hypothesis to account for the whole supply of solar heat; but there is as yet no direct evidence whatever that the sun is really shrinking.

The basis of the meteoric theory is simply this: If a moving body be stopped, either suddenly or gradually, a quantity of heat is generated which may be expressed, in calories, by the formula $\frac{mv^2}{850}$, in which m is the mass of the body, in kilogrammes, and v its velocity, in metres per second. A body weighing 850 kilogrammes, and moving one metre per second, would, if stopped, develop just one calory of heat—i. e., enough to heat one kilogramme of water from freezing-point to 1° Cent. If it were moving five hundred metres per second (about the speed of a cannon-ball), it would produce two hundred and fifty thousand times as much heat, or enough to raise the temperature of a mass of water equal to itself nearly 300° Cent. If it were moving, not five hundred metres per second, but about seven hundred thousand (approximately the velocity with which a body would fall into the sun from any planetary distance), the heat produced would be 1,400 × 1,400, or nearly two million, times as great—sufficient to bring a mass of matter many thousand times greater than itself to most vivid incandescence, and immensely more than could be produced by its complete combustion under any conceivable circumstances. With reference to this theory, Sir William Thomson has calculated the amount of heat which would be produced by each of the planets in falling into the sun from its present orbit. The results are as follows, the heat produced being expressed by the number of years and days through which it would maintain the sun's present expenditure of energy:

	Years.	Days.
Mercury	6	219
Venus	83	326
Earth	95	19
Mars	12	259
Jupiter	32,254	
Saturn	9,652	
Uranus	1,610	
Neptune	1,890	
Total	45,604	

That is, the collapse of all the planets upon the sun would generate sufficient heat to maintain its supply for nearly forty-six thousand years. A quantity of matter equal to only about one one-hundredth of the mass of the earth, falling annually upon the solar surface, would, therefore, maintain its radiation indefinitely. Of course, this increase of the sun would cause an acceleration of the motion of all the planets—a shortening of their periods. Since, however, the mass of the sun is three hundred and thirty thousand times that of the earth, the yearly addition would be only one thirty-three-millionth of the whole, and it would require centuries to make the effect sensible. The only question, then, is, whether any such quantity of matter can be supposed to reach the sun. While it is impossible to deny this dogmatically, it, on the whole, seems improbable, for astronomical reasons. In the first place, if meteoric matter is so abundant, the earth ought to encounter much more of it than she does; enough, in fact, to raise her temperature above that of boiling water.

Then, again, if so large a quantity of matter annually falls upon the solar surface, it is necessary to suppose a vastly greater quantity circulating around the sun between it and the planet Mercury. The process by which the orbit of a meteoric body is so changed as to make it enter the solar atmosphere is a very slow one, so that only a very small proportion of the whole could be caught in any given year. Now, if there were near the sun any considerable quantity of meteoric matter—anything like the mass of the earth, for instance—it ought to produce a very observable effect upon the motions of the planet Mercury, an effect not yet detected.* For this reason astronomers generally, while conceding that a portion, and possibly a considerable fraction, of the solar heat may be accounted for by this hypothesis, are disposed to look further for their explanation of the principal revenue of solar energy. They find it in the probable slow contraction of the sun's diameter, and the gradual liquefaction and solidification of the gaseous mass. The same total amount of heat is produced when a body moves against a resistance which brings it to rest gradually as if it had fallen through the same distance freely and been suddenly stopped. If, then, the sun does contract, heat is necessarily produced by the process, and that in enormous quantity, since the attracting force at the solar surface is more than twenty-seven times as great as gravity at the surface of the earth, and the contracting mass is so immense.

* Leverrier considered that he had detected in the motions of Mercury an irregularity of the kind indicated, but much smaller. It was such, according to his calculations, as would be accounted for by the action of one or several planets whose aggregate mass should be much less than that of the earth. This was the basis on which he founded his strong belief in the existence of the intra-Mercurial planet Vulcan.

In this process of contraction, each particle at the surface moves inward by an amount equal to the whole diminution of the solar radius, while a particle below the surface moves less, and under a diminished gravitating force; but every particle in the whole mass of the sun, excepting only that at the exact center of the globe, contributes something to the evolution of heat. To calculate the precise amount of heat developed, it would be necessary to know the law of increase of the sun's density from the surface to the center; but Helmholtz, who first suggested the hypothesis, in 1853, has shown that, under the most unfavorable suppositions, a contraction in the sun's diameter of about two hundred and fifty feet a year—a mile in a trifle over twenty-one years—would account for its whole annual heat-emission. This contraction is so slow that it would be quite imperceptible to observation. It would require nine thousand five hundred years to reduce the diameter a single second of arc (since $1''$ equals 450 miles at the sun's distance), and nothing less would be certainly detectible.

Of course, if the contraction is more rapid than this, the mean temperature of the sun must be actually rising, notwithstanding the amount of heat it is losing. Observation alone can determine whether this is so or not.

If the sun were wholly gaseous, we could assert positively that it must be growing hotter; for it is a most curious (and at first sight paradoxical) fact, first pointed out by Lane in 1870, that the temperature of a gaseous body continually rises as it contracts from loss of heat. By losing heat it contracts, but the heat generated by the contraction is more than sufficient to keep the temperature from falling. A gaseous mass losing heat by radiation, must, therefore, at the same time grow both smaller and hotter, until the density becomes so great that the ordinary laws of gaseous expansion reach their limit and condensation into the liquid form begins. The sun seems to have arrived at this point, if indeed it were ever wholly gaseous, which is questionable. At any rate, so far as we can now make out, the exterior portion—i. e., the photosphere—appears to be a shell of cloudy matter, precipitated from the vapors which make up the principal mass, and the progressive contraction, if it is indeed a fact, must result in a continual thickening of this shell and the increase of the cloud-like portion of the solar mass.

This change from the gaseous to the liquid form must also be accompanied by the liberation of an enormous quantity of heat, sufficient to materially diminish the amount of contraction needed to maintain the solar radiation.

Of course, if this theory of the source of the solar heat is correct, it follows that in time it must come to an end; and looking backward we see that there must also have been a beginning. Time was when there was no such solar heat as now, and the time must come when it will cease.

We do not know enough about the amount of solid and liquid matter at present in the sun, or of the nature of this matter, to calculate the future duration of the sun with great exactness, though an approximate estimate can be made. The problem is a little complicated, even on the simplest hypothesis of purely gaseous contraction, because as the sun shrinks the force of gravity increases, and the amount of contraction necessary to generate a given amount of heat becomes less and less; but this difficulty is easily met by a skillful mathematícian. According to Newcomb, if the sun maintains its present radiation it will have shrunk to half its present diameter in about five million years at the longest. As it must, when reduced to this size, be eight times as dense as now, it can hardly then continue to be mainly gaseous, and its temperature must have begun to fall. Newcomb's conclusion, therefore, is that it is hardly likely that the sun can continue to give sufficient heat to support life on the earth (such life as we now are acquainted with, at least) for ten million years from the present time.

It is possible to compute the past of the solar history upon this hypothesis somewhat more definitely than the future. The present rate of contraction being known, and the law of variation, it becomes a purely mathematical problem to compute the dimensions of the sun at any date in the past, supposing its heat-radiation to have remained unchanged. Indeed, it is not even necessary to know anything more than the present amount of radiation, and the mass of the sun, to compute how long the solar fire can have been maintained, at its present intensity, by the process of condensation. No conclusion of geometry is more certain than that the contraction of the sun from a diameter even many times larger than that of Neptune's orbit to its present dimensions, if such a contraction has actually taken place, has furnished about eighteen million times as much heat as the sun now supplies in a year; and therefore that the sun can not have been emitting heat at the present rate for more than that length of time, if its heat has really been generated in this manner. If it could be shown that the sun has been shining as now for a longer time than that, the theory would be refuted; but if the hypothesis be true. as it probably is in the main, we are inexorably shut up to the conclusion that the total life of the solar system, from its birth to its death, is included in some such space of time as thirty million years. No reasonable allowances for the fall of meteoric matter, based on what we are now able to observe, or for the development of heat by liquefaction, solidification, and chemical combination of dissociated vapors, could raise it to sixty million.

At the same time, it is of course impossible to assert that there has been no catastrophe in the past—no collision with some wandering star, endued, as Croll has supposed, like some of those we know of now in the heavens, with a velocity far surpassing that to be acquired by a fall even from infinity, producing a shock which might in a few hours, or moments even, restore the wasted energy of ages. Neither is it wholly safe to assume that there may not be ways, of which we yet have no conception, by which the energy apparently lost in space may be returned, and burned-out suns and run-down systems restored; or, if not restored themselves, be made the germs and material of new ones to replace the old.

But the whole course and tendency of Nature, so far as science now makes out, points backward to a beginning and forward to an end. The present order of things seems to be bounded, both in the past and in the future, by terminal catastrophes, which are veiled in clouds as yet impenetrable.

CHAPTER IX.

SUMMARY OF FACTS, AND DISCUSSION OF THE CONSTITUTION OF THE SUN.

Table of Numerical Data.—Constitution of Sun's Nucleus.—Peculiar Properties of Gases under High Temperature and Pressure.—Characteristic Differences between a Liquid and a Gas.—Constitution of the Photosphere and Higher Regions of the Sun's Atmosphere.—Professor Hastings's Theory.—Pending Problems of Solar Physics.

It may be well to collect into a brief summary the principal facts and conclusions of the preceding pages, presenting them in a single comprehensive view. We give first, therefore, a table of the *statistics* of the sun —the facts which can be stated in numbers:

Solar parallax (equatorial horizontal), $8.80'' \pm 0.02''$.
Mean distance of the sun from the earth, 92,885,000 miles; 149,480,000 kilometres.
Variation of the distance of the sun from the earth between January and June, 3,100,000 miles; 4,950,000 kilometres.
Linear value of $1''$ on the sun's surface, 450.3 miles; 724.7 kilometres.
Mean angular semidiameter of the sun, $16' 02.0'' \pm 1.0''$.
Sun's linear diameter, 866,400 miles; 1,394,300 kilometres. (This may, perhaps, be *variable* to the extent of several hundred miles.)
Ratio of the sun's diameter to the earth's, 109.3.
Surface of the sun compared with the earth, 11,940.
Volume, or cubic contents, of the sun compared with the earth, 1,305,000.
Mass, or quantity of matter, of the sun compared with the earth, $330,000 \pm 3,000$.
Mean density of the sun compared with the earth, 0.253.

SUMMARY OF FACTS, ETC. 279

Mean density of the sun compared with water, 1·406.
Force of gravity on the sun's surface compared with that on the earth, 27·6.
Distance a body would fall in one second, 444·4 feet; 135·5 metres.
Inclination of the sun's axis to the ecliptic, 7° 15'.
Longitude of its ascending node, 74°.
Date when the sun is at the node, June 4–5.
Mean time of the sun's rotation (Carrington), 25·38 days.
Time of rotation of the sun's equator, 25 days.
Time of rotation at latitude 20°, 25·75 days.
Time of rotation at latitude 30°, 26·5 days.
Time of rotation at latitude 45°, 27·5 days.

(These last four numbers are somewhat doubtful, the formulæ of various authorities giving results differing by several hours in some cases.)

Linear velocity of the sun's rotation at his equator, 1·261 miles per second; 2·028 kilometres per second.
Total quantity of sunlight, 6,300,000,000,000,000,000,000,000 candles.
Intensity of the sunlight at the surface of the sun, 190,000 times that of a candle-flame; 5,300 times that of metal in a Bessemer converter; 146 times that of a calcium-light; 3·4 times that of an electric arc.
Brightness of a point on the sun's limb compared with that of a point near the center of the disk, 25 per cent.
Heat received per minute from the sun upon a square metre, perpendicularly exposed to the solar radiation, at the upper surface of the earth's atmosphere (*the solar constant*), 25 calories.
Heat-radiation at the surface of the sun, per square metre per minute, 1,117,000 calories.
Thickness of a shell of ice which would be melted from the surface of the sun per minute, 48¼ feet; or 14¾ metres.
Mechanical equivalent of the solar radiation at the sun's surface, continuously acting, 109,000 horse power per square metre; or 10,000 (nearly) per square foot.
Effective temperature of the solar surface (according to Rosetti), about 10,000° Cent.; or 18,000° Fahr.

Of course, it hardly need be repeated here that the figures relating to the light and heat of the sun are much less reliable than those which refer to its distance, dimensions, mass, and attracting power.

The cut on page 281 is intended to present to the eye, more clearly than any mere description could do, the constitution of the sun, and the relation of the different concentric shells or envelopes of which it is formed.

The picture is an ideal section through the center. The black disk represents the inner nucleus, which is not accessible to observation, its nature and constitution being a mere matter of inference. The white ring surrounding it is the photosphere, or shell of incandescent cloud which forms the visible surface. The depth, or thickness, of this shell is quite unknown; it may be many times thicker than represented, or possibly somewhat thinner. Nor is it certain whether it is separated from the inner core by a definite surface, or whether, on the other hand, there is no distinct boundary between them.

The outer surface of the photosphere, however, is certainly pretty sharply defined, though very irregular, rising at points into faculæ, and depressed at others in spots, as shown in the figure.

Immediately above this lies the so-called "reversing stratum," in which the Fraunhofer lines originate. It is to be noted, however, that the gases which compose this stratum do not merely *overlie* the photosphere, but they also fill the interspaces between the photospheric clouds, forming the atmosphere in which they float, and an attempt has been made to indicate this fact in the diagram.

Above the "reversing stratum" lies the scarlet chromosphere, with prominences of various forms and dimensions rising high above the solar surface; and over, and embracing all, is the coronal atmosphere and the mysterious radiance of clouds, rifts, and streamers, fading gradually into the outer darkness.

At the center of the sun the earth is represented in its true relative dimensions—$\frac{1}{109}$ of the three inches

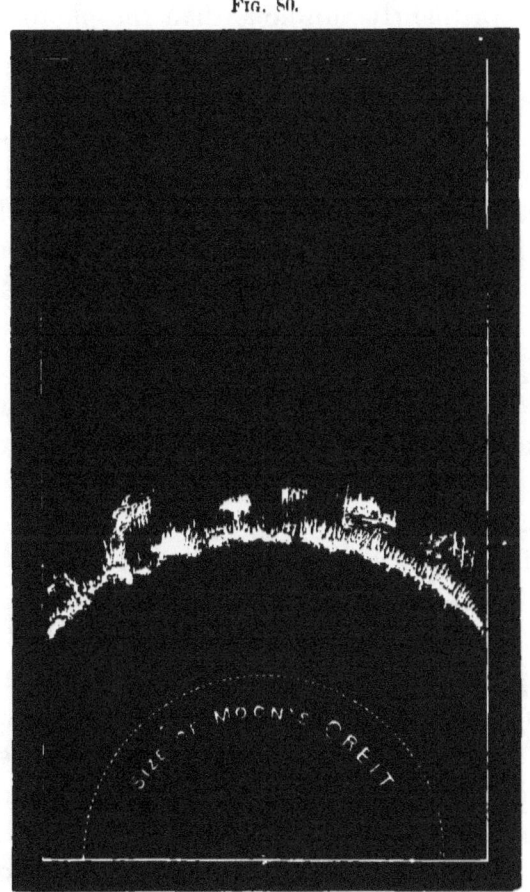

Fig. 80.

which is taken as the scale of the sun's diameter. This scale reduces our globe to a little dot only $\frac{1}{36}$ of an inch across. Around it, at its proper distance, is drawn the orbit of the moon, still far within the photosphere, the moon herself being fairly represented by any one of the minute points which make up the dotted line that indicates her path.

The central nucleus is made black in the picture, simply for convenience, and not with any purpose to indicate that the matter which composes it is cooler or even less brilliantly luminous than the photosphere. It is quite probable, indeed, that this central core (which contains certainly more than nine tenths of the whole mass of the sun) is purely gaseous, and it is of course true that, *at a given temperature and pressure*, a gaseous mass has a lower radiating power, and is less luminous, than a mass of clouds, such as those which constitute the photosphere. But, on the other hand, both compression and increase of temperature rapidly raise the radiating power of a gas; and it is highly probable that, at no very considerable depth, the growing pressure and heat may more than equalize matters, and render the central nucleus as intensely bright as the surface of the sun itself.

At the upper surface of the photosphere, however, and all through it, indeed, the uncondensed gases are dark as compared with the droplets and crystals which make up the photospheric clouds. Here the pressure and temperature are lowered, so that the vapors give out no longer a continuous but a bright-line spectrum, whenever we get a chance to see them, against a non-luminous background; and, when the intenser light from the liquid and solid particles of the photosphere shines through these vapors, they rob it of the corresponding rays, and produce for us the familiar dark-lined spectrum of ordinary sunlight.

It is, perhaps, hardly necessary to state again the reasons for believing the great body of the sun to be gaseous; the argument depends upon the enormous heat at the surface, which keeps the solar atmosphere charged with the vapors of our familiar metals, and the fact that the mean density of the sun is so low (only one and one fourth times that of water), that it is quite impossible that any of the substances which we have reason to believe to exist in the sun could have the solid, or even the liquid, form through any considerable portion of its mass. That is to say, if any large proportion of the whole were composed of solid or liquid iron, titanium, magnesium, etc., the density would be far greater than it really is; and, since the temperature, at the surface even, where there is free radiation and exposure to the cold of space, is so high as to keep these bodies in the state of vapor, it is not likely that, at greater depths, it is low enough to permit their liquefaction or solidification.

And yet the theory that they are in a gaseous state is not free from difficulties. A few years ago it would have been urged with great plausibility that, under the enormous pressure due to the weight of the superincumbent mass acted upon by the solar gravity—nearly twenty-eight times that of the earth, it is to be remembered—any gas whatever must be liquefied at no very great depth below the surface.

Even on the earth, for example, the density of the air decreases one half for every three and a half miles of elevation, and it ought to increase in a similar proportion for every three and a half miles of descent below the sea-level, if we drop for a moment considerations relative to temperature. Since water is about seven hundred and seventy times as heavy as the air at the earth's surface, it follows, therefore, that at the bottom of a shaft thirty-five miles deep the air would be more dense than water, if of the same temperature as at the surface; and, before a depth of fifty miles were reached, it would become denser than gold, unless it had first liquefied, and so become less compressible. If we take account of the slight decrease of the force of gravity as we go below the earth's surface, and assume that the temperature increases, even at the rate of 100° Fahr. for each mile of descent, the results will be modified, but not materially changed in character. It would merely be necessary to go some ten miles deeper to reach the same result.

Now, at the sun, where the action of gravity is so much more intense, it is evident that, *unless the temperature rises very rapidly below the surface*, or unless liquefaction supervenes, the density of gases must increase so fast that the mean density of the mass—if the sun be really gaseous—must be vastly greater than that of any known metal.

But liquefaction, as we now know, can not take place under the circumstances. The researches of Andrews and others have shown that to effect the liquefaction of a gas two things must go together—increase of pressure and diminution of temperature. For each gas there is a so-called "critical temperature," and, so long as the temperature does not fall below this point, no pressure whatever can reduce the gas to the liquid form. When the temperature has fallen below it, then pressure alone will produce the desired effect, and, if the temperature is very low, only a slight degree of pressure will be needed. Now on, or in, the sun the temperature can not be supposed to be below the "critical points" of several of the gases found there, and hence, as has been said, their liquefaction is out of the question. Those, therefore, who are unwilling to admit a sufficient increase of temperature with increasing depth below the solar surface, have been disposed to hold that the central portions of the sun are not composed, to any great extent, of the same elements which the spectroscope reveals to us in the solar atmosphere, but of some different unknown solid or liquid substance, of great rigidity and low density. With this view, generally, also goes the belief that the evolution of solar heat is essentially a surface-action, produced, by some unexplained process, only where the exterior of the solar orb encounters open space, and not of necessity implying any great heat in the inner depths. The older observers, especially the Herschels, for the most part held theories essentially like that sketched above. The elder Herschel, it will be remembered, even contended pretty vigorously that the central globe of the sun is a habitable world, sheltered from the blazing photosphere by a layer of cool, non-luminous clouds. And in more recent times Kirchhoff and Zöllner have maintained that the luminous surface is either liquid or solid.

While it is, perhaps, not possible to demonstrate at present the falsity of this theory, by proving that the solar nucleus is neither solid nor liquid, and showing that the solar heat is not confined to the surface, but permeates the whole mass with continually increasing intensity near the center of the globe, it is yet evident enough that it meets the exigencies of the case only by calling in unknown and imaginary substances and operations. On the other hand, the gaseous theory, which is now generally adopted, involves no new kinds of matter or unknown forces, but conceives of solar phenomena as entirely the same in kind as those we are familiar with in our laboratories, though immensely different in degree and intensity.

If we only grant that the temperature rises rapidly

enough from the surface downward through the solar globe, the whole difficulty as to the density of such a gaseous sphere vanishes. It is true that, on this view, the central temperature must be tremendous, even in comparison with that of the photosphere. But why not? Can any reason be assigned to the contrary? If we could suppose the sun wholly made of hydrogen, and that the ordinary relations deduced by our laboratory experiments hold between the pressure and temperature through all possible ranges of both, it would then be a comparatively simple matter to compute the least central temperature which would give the solar globe its present density. If, however, we remember that other materials, and in unknown proportions, enter into the problem, and that in all probability our laboratory-work gives only approximate formulæ, it is clear that such a computation would be useless. We must content ourselves for the present with vague expressions, and say roughly that the intensity of the sun's internal heat must as much exceed that of the photosphere as this surpasses the mere animal warmth of a living body.

But while, on the whole, it thus seems probable that the sun's core is gaseous, nothing could be remoter from the truth than to imagine that a mass of gas, under such conditions of temperature and pressure, would resemble our air in its obvious characteristics. It would be denser than water; and since, as Maxwell and others have shown, the viscosity of a gas increases fast with rising temperature, it is probable that it would resist motion something like a mass of pitch or putty.

One might, then, naturally enough ask, why a substance so widely different from gases as we know them by experience, and so much resembling what we are accustomed to call the semifluids, should not be classed with them rather than with the gases. The reply, of course, is, that although the substance thus bears a superficial likeness to the semifluids, its essential characteristics are still those of a gas, viz., continuous expansion under diminishing pressure without the formation of a free surface of equilibrium; continuous expansion under increasing temperature without the attainment of a boiling-point; and, in the case of a mixture of different gases, a uniform diffusion of each, according to Dalton's law, without regard to specific gravity.

Perhaps a little fuller explanation may be allowed on this point, which is often misunderstood. Suppose a mass of liquid to be contained in a close vessel, which it just fills, and compressed by some enormous force; now let the vessel grow gradually larger, thus relieving the pressure. The liquid will expand, at first keeping the vessel full; but at last, even if heat be supplied to prevent the temperature from falling, a time will come when the liquid will no longer fill the vessel, but an empty space will be left above a well-defined "free surface of equilibrium"—a space empty, that is, of the liquid, but of course occupied by its vapor. Now, if we take a similar vessel filled with a compressed gas, the density of which may, on account of the pressure, at first even exceed that of the liquid in the case just cited, and allow the vessel to expand in the manner described, at the same time supplying heat enough to keep the temperature from falling, the gas will never cease to fill the whole vessel, nor will it ever form a free surface like the liquid, however far the enlargement of the vessel may be carried.

Again, if we take a cylinder with a weighted piston, fitting it and moving freely in it, and, after filling the space below the piston with a liquid, apply heat to it, we shall find that at first the temperature will rise regularly, and the liquid, expanding slightly as it warms, will push the piston before it. But, when a certain temperature, depending upon the nature of the liquid and the pressure exerted by the piston, has been reached, the liquid will cease to grow hotter by the further application of heat, and will begin to boil; and the liberated vapor will raise the piston and occupy the otherwise vacant space above the surface of the liquid. If, however, the space originally below the piston were occupied by a gas, however dense, no such thing would happen. The gas, on the application of heat, would rise in temperature, and expand regularly without discontinuity or limit.

Finally, as to the third criterion which marks the difference between liquids and gases. In a mixture of liquids of different specific gravities, the different materials separate and arrange themselves in strata, according to their weights, unless they have some chemical action on each other—for example, quicksilver, water, and oil. But a mixture of several gases, differing however widely in specific gravity—for example, hydrogen, oxygen, and carbon dioxide—behaves in no such way: under all conditions of temperature and pressure each gas distributes itself through the whole space, precisely as if the others were not present, only more slowly than if it were alone.

Although it may not be possible, in the present state of science, to demonstrate that the principal portion of the solar mas is gaseous, this much can at least be said—that a globe of incandescent gas, under conditions such as have been intimated, would necessarily present just such phenomena as the sun exhibits.

On the outer surface, exposed to the cold of space,

the rapid radiation would certainly produce the condensation and precipitation into luminous clouds of such vapors as had a boiling-point higher than that of the cooling surface. These clouds would float in an atmosphere saturated with the vapors from which they were formed, and also containing such other vapors as were not condensed, and thus the peculiarities of the solar spectrum would result. On the other hand, the permanent gases, like hydrogen—those not subject to condensation into the liquid form under the solar conditions—would rise to higher elevations than the others, and form above the photosphere just such a chromosphere as we observe. Whether, from the mere assumption of such a constitution for the sun, one could work out, *a priori*, the phenomena of sun-spots and prominences, is indeed doubtful; but thus far nothing in any of them has been observed which appears to be inconsistent with this view of the subject —nothing, we say, unless it should turn out, as some observers suspect, that the solar surface possesses, so to speak, " geographical " characteristics, evinced by the disposition to break out into sun-spots at certain fixed points—as if at those points there were volcanoes or something of the sort. Of course, the fact that the spots are distributed mainly in two belts parallel to the solar equator, involves no difficulty, for it is easy to conceive how, in more than one way, the sun's rotation might lead to such a result: but peculiarities permanently attaching to individual points on the solar surface necessarily imply rigid connections, such as are inconsistent with the theory of a gaseous or even of a fluid nucleus. At present, astronomers generally are not disposed to admit that such fixed " spot-centers " exist; and yet considerable weight is certainly due to the opinion of so experienced an observer as Spoerer, who seems to favor the idea. At first sight, it would appear as if the question might be easily settled by reference to any extended series of observations, like those of Schwabe or Carrington. But, if there really is such a solid nucleus, its time of rotation is unknown, and this makes the discussion difficult and unsatisfactory. On the whole, the weight of evidence is heavily in favor of the received theory.

With reference to the constitution of the photosphere there is a general agreement among astronomers. A few, indeed, still hold, as has been mentioned, to the idea that the visible surface is a liquid sheet; but the whole appearance of things, the details of the granulation, the phenomena of spots and faculæ, the mobility and variability of the floccules, all better accord with the theory adopted in these pages, which is a necessary consequence of the hypothesis that the sun is principally gaseous. It seems almost impossible to doubt that the photosphere is a shell of clouds. As to the precise constitution of this shell, however, the form and magnitude of the component cloudlets, the chemical elements involved, and the temperature and pressure, there is room for a good deal of uncertainty and difference of opinion. The more common view, apparently—the one, certainly, which the writer has hitherto held—is, that the clouds are formed mainly by the condensation of the substances which are most conspicuous in the solar spectrum, such as iron and the other metals. As to the form of the clouds, also, it has usually been assumed that, as a consequence of the ascending currents by which they are formed, they are columnar, their height being much greater than their other dimensions.

Professor Hastings, of Baltimore, has recently published a somewhat different theory (already referred to in a previous chapter), which has much to recommend it, and avoids some of the difficulties of the received doctrine, though not without encountering others which seem, at first sight, just as formidable. We can not do better than to quote the concluding page of his paper, published in the "Proceedings of the American Academy of Arts and Sciences" (Boston, November, 1880), and in the "American Journal of Science" for January, 1881:

"The theory of the constitution of the sun above proposed may be recapitulated as follows: Convection currents, directed generally from the center of the sun, start from a lower level, where the temperature is probably above the vaporizing temperature of every substance. As these currents move upward they are cooled mainly by expansion, until a certain element (probably of the carbon group) is precipitated. This precipitation, restricted from the nature of the case, forms the well-known granules. There is nothing which has come under my observation which would indicate a columnar form in these granules, under ordinary circumstances."

The main peculiarity of the hypothesis, thus far, consists in the idea, stated in an earlier part of his paper, that the photospheric "clouds" are formed by the precipitation of either carbon, silicon, or boron (the three members of the carbon group), to the exclusion of other substances which are less refractory (have *lower boiling-points*), and therefore escape precipitation. Those bodies which have boiling-points higher than that of this photosphere-element, as it may be called, will, therefore, not exist to any extent in the vaporous atmosphere, having suffered precipitation before they reach the visible surface. Those only will show their lines in the spectrum which have lower boiling-points, and so do not suffer precipitation at the temperature of the photosphere. This is the reason why the lines of silicon, etc., do not appear in the solar spectrum, while those of iron, etc., do. Of course, it will at once be seen that, if this view is true, the temperature of the photosphere is that of the boiling-point (under the local conditions of pressure) of the silicon or carbon, or whatever it is which forms the clouds. As an objection to the view, it immediately occurs to one that, if the carbon, for instance, *is* precipitated at and below some special elevation, yet the iron vapor will rise above it, and, in its turn, will find a level and temperature of precipitation, so that the photospheric clouds, instead of being composed of any single substance, would contain all which can find a level and temperature of precipitation anywhere in the solar atmosphere. As to the form of the floccules, it would seem that the successive precipitation, at different levels and temperatures of different elements in an ascending current, must result in clouds of great vertical extent.

But to resume the quotation :

"The precipitated material rapidly cools on account of its great radiating power, and forms a fog or smoke which settles slowly through the spaces between the granules, until revolatilized below. It is this smoke which produces the general absorption at the limb, and the 'rice-grain' structure of the photosphere.

"Where any disturbance tends to increase a downward convection current, there is a rush of vapors at the outer surface of the photosphere toward this point. These horizontal currents or winds carry with them the cooled products of precipitation, which, accumulating above, dissolve slowly below in sinking. This body of smoke forms the solar spot.

"The upward convection currents in the region of the spots are bent horizontally by the centripetal winds. Yielding their heat now, by the relatively slow process of radiation, the *loci* of precipitation are much elongated, thus giving the region immediately surrounding a spot the characteristic radial structure of the penumbra.

"This conception of the nature of the penumbra implies a ready interpretation of a remarkable phenomenon, amply attested by the most skillful observers, and, as far as my knowledge goes, wholly unexplained, namely, the brightening of the inner edge of the penumbra in every well-developed spot.

"This interpretation is, perhaps, most readily imparted by a comparison of the hot convection currents in the two cases. When the convection current is rising vertically, the medium is cooled by expansion until the precipitation temperature is reached, when all the condensible material appears *suddenly*, save as it is somewhat retarded by the heat liberated in the act. Immediately afterward the particles become relatively dark by radiation. In the horizontal currents a very different condition of things obtains. Here the medium does not cool dynamically, by expansion, but only by radiation; hence, since the radiation of the solid particles is enormously greater than that of the supporting gas, practically by that of the particles themselves. Thus, after the first particle appears, it must remain at its brightest incandescence until all the material of which it is composed is precipitated. From this we see that such an horizontal current must increase gradually in brilliancy to its maximum, and then suddenly diminish—an exact accordance with the facts as observed."

The idea that the stratum which produces the general absorption at the limb of the sun is a veil of "smoke"—i. e., of the same minute particles which constitute the photosphere, but cooled to relative darkness—has been already alluded to in a preceding chapter. So far as we know, it is novel and valuable, clearing up a good many embarrassing difficulties. It is so obvious, on reflection, that something of the sort must accompany the photosphere, that it is surprising that the idea has not been thought of before. Of course, the particles formed by condensation must, many of them at least, be carried by the ascending currents high above the point of their formation, and cooled so much as to become relatively dark in comparison with the more vivid incandescence of the regions below, just as the ascending particles of carbon, unconsumed and cooled, constitute the smoke of a fire. As regards the explanation of spot phenomena, we see no special advantage, or indeed novelty, in the idea proposed. The received theory regards the general brightening at the inner edge of the penumbra as produced by the convergence of the luminous filaments, rendered horizontal by the indraught. The *quasi*-bulbous termination of the filaments occurs only occasionally, and may, perhaps, be accounted for in the way proposed by Mr. Hastings more satisfactorily than in any other; still, many circumstances seem to indicate that the brightening at the end is due, like that of the faculæ, to mere protrusion through the smoke-veil.

As regards the chromosphere and "reversing stratum" very little needs to be added. Perhaps a caution may be in place, that they and the vapors of the photosphere are not to be thought of as entirely separate and distinct. *All* the gases are found together in the interstices between the cloud-granules of the photosphere— the unknown substance which produces the green line in the spectrum of the corona, the hydrogen and hypothetical "helium" which characterize the chromosphere, and the metallic vapors which give the reversing layer its peculiar properties—these all exist together in the lower depths, unless, indeed, it may possibly be the case that at the greater elevations some compound bodies are formed which can not exist in the fiercer fires below. So far as we can distinguish between these different portions, we may define the photosphere as the shell within which precipitation is taking place; the reversing layer, as that lowest region of the solar atmosphere which contains sensibly all the gases which the spectroscope indicates to us; the chromosphere, as the region of hydrogen and "helium"; and the corona, as that upper domain of the solar atmosphere which becomes observable only during solar eclipses. But the coronal gas itself is most conspicuous and abundant right in the photosphere and reversing layer, and the same is true of the hydrogen of the prominences.

It is well, also, to bear in mind that, if any substances decomposable by heat exist upon the sun at all, we must expect to find them in the higher and cooler regions of the solar atmosphere. In and near the photosphere, or underneath it, matter must be in its most elemental state.

As to the mechanism of the chromosphere and prominences, if we may use the expression, much certainly remains to be learned. In many cases, indeed, perhaps in most, the forms and behavior of the protuberances are satisfactorily enough accounted for by supposing that the heated hydrogen and its associate vapors is simply forced up into cooler regions by pressure from below—a pressure which must result from the downward movement of the great mass of precipitated matter which forms the photosphere. But evidently this is not the whole story. We must have recourse to ideas of a different order to account for the somewhat rare, but still really numerous and well-authenticated, instances when the summits of prominences have been seen to rise in a few minutes to elevations of two or three hundred thousand miles, the upward motion being almost visible to the eye at the rate of a hundred miles a second or more.

Very perplexing, also, is the indubitable fact that

clouds of this prominence-matter sometimes gather and form without any apparent connection with the chromosphere below, apparently just as clouds form in our own atmosphere, by the condensation of vapor before invisible. On the whole, it looks very much as if we must regard the prominences as differing from the surrounding medium mainly, if not wholly, in their luminosity —as simply superheated portions of an immense atmosphere.

But, then, we immediately encounter the difficulties so ably urged by Lane, Lockyer, and others, that the existence of hydrogen of any appreciable density, at the elevation of even a hundred thousand miles, implies a density and pressure at the surface of the photosphere so high as to be entirely inconsistent with the spectroscopic phenomena there manifested—unless, indeed, under solar conditions, the action of gravity upon the gases of the solar atmosphere is modified by some repulsive force. That such a force is at least conceivable, is obvious from the behavior of the tails of comets; and many features in the corona point in the same direction. Of its nature and origin we can not, however, assert anything as yet.

Even more difficult than the problem of the chromosphere is that of the corona. While it is something to know that the phenomenon is mainly solar, and that, therefore, it must rank in magnitude and importance with the most magnificent of natural objects, we have yet to find a satisfactory explanation of many of its most obvious features. It is certainly very complex— matter meteoric and matter truly solar; orbital motion, solar attraction, atmospheric resistance, and actions thermal, electrical, and magnetic, are probably all combined.

At present it would seem that the most important and fundamental problems of solar physics which are now pressing for solution are these : first, a satisfactory explanation of the peculiar law of rotation of the sun's surface ; second, an explanation of the periodicity of the spots, and their distribution ; third, a determination of the variations in the amount of the solar radiation at different times and different points upon its surface; and, fourth, a satisfactory explanation of the relations of the gases and other matters above the photosphere to the sun itself—the problem of the corona and the prominences.

One might name many others of hardly less interest, such as that which has to do with the intimate connection between terrestrial magnetism and the condition of the solar surface ; but, on the whole, the four named seem to be those the solution of which would most advance our science. Not, of course, that we are to suppose that even their solution would bring us in sight of the end or limit of knowledge. Each onward step only opens before us a new, wider, and more magnificent horizon, with infinity still beyond it.

APPENDIX.

PROFESSOR LANGLEY'S ACCOUNT OF HIS BOLOMETRIC OBSERVATIONS, AND CERTAIN CONCLUSIONS DERIVABLE FROM THEM.

The author of this work has done me the honor to ask me to write some notice of a research, on which I am now engaged, in part supplementary to what is given in the text. I do so with much pleasure, but with the request that the reader will remember that what he now reads has not yet become part of the body of accepted scientific fact, but rests on my own statement of opinion. He should also observe that it is even thus given, not as exact but only as approximate truth.

Although it is well understood that the expressions "thermal," "luminous," and "chemical" rays are misleading, and that we are, in the solar spectrum, concerned only with one and the same energy, which is interpreted to us in terms of heat, light, and chemical action, according to the medium by which it is perceived, an experimental proof of this can hardly be without interest.

Again, even if we adopt without reserve the doctrine just repeated, we are still in danger of using the diagrams of the text-books, where three curves are given for "heat," "light," and "actinism," under the extremely general misapprehension that these curves do at least show the distribution of the energy in the spectrum with approximate truth.

No *single* complete and satisfactory experimental proof can, perhaps, ever be given, till we produce our spectrum by a medium which has no selective absorption whatever, and by a medium which can also be so used as to give a normal distribution of the energy. We can, however, by means of the diffraction reflecting grating, form a nearly normal spectrum, in which the energy is approximately so distributed, and in which the selective absorption is slight, relatively, to that in the prism. When this is done, a consideration of the result will remove the misapprehension just mentioned as to the distribution of the solar energy, or will at least give us much more just ideas of what it really is.

The reason that this has not been done long ago is not on account of any failure to recognize its desirability, but owing to the difficulty, amounting nearly to impossibility, of making the measurement in detail, so as to show the relative energy in different parts, that is, how much really inheres in the visible portion and how much in the invisible. We have now, for example, familiar diagrams showing that the greatest solar energy is found in rays lying in the ultra-red, and that the portion of energy which is employed in making us *see* is not greatly different in amount from that which is found in the ultra-violet radiations. Conclusions of the first kind have a wide bearing on most important meteorological questions, if on no other. Those of the second kind are also of consequence, and both are, as it seems to me, erroneous.

As the heat in the diffraction spectrum is, at best,

about one tenth that in the prismatic—which is itself all but immeasurably small when distributed in approximately homogeneous rays—special apparatus has been devised for the peculiarly delicate measurements in the diffraction spectrum, which I have lately succeeded in making. The apparatus depends on the principle (not in itself at all new) that, if, of two wires from a battery, making the arms of an electric "bridge," or "balance," we warm only one, a galvanometer needle may be made to move, owing to the diminished current caused by the heat. But, though the principle is simple, the special application has been difficult. The instrument, as finally constructed for measuring most minute portions of radiant energy, as heat, uses strips of metal about $\frac{1}{100000}$ inch thick as the balance-arms, and I have called it the *Bolometer*.* With the one I am now using, a change of temperature of about 0·00001° Cent. in the strips is detected, a change of $\frac{1}{100000}$ degree being noted instantly. As these strips are extremely minute, this implies a power of recognizing amounts of radiant heat smaller than those for which the thermopile is commonly employed. *How* small it is difficult to apprehend clearly, but it may be stated, in illustration both of the feebleness of radiant energy in some parts of the diffraction spectrum and of the delicacy of the instrument, that the heat in certain ultra-violet rays can be detected by it in rather less than ten seconds, though the same radiation is so weak that, falling uninterruptedly for over one thousand years on a kilogramme of ice at 0° Cent., it would not wholly melt it.

With this apparatus, measuring approximately *homo-*

---

* I desire to mention that the cost of the experimental construction of the Bolometer has been principally met by aid from the Rumford fund, through the American Academy of Arts and Sciences.

APPENDIX. 301

geneous rays ; of wave-lengths represented by the numbers on the horizontal line, and of energies corresponding for the particular wave-length to the perpendicular, we obtain the following curve (at Allegheny, with winter sun), which represents at least, with a rough approxi-

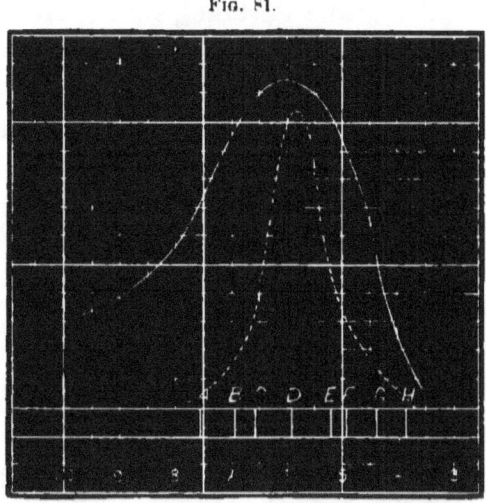

Fig. 81.

The full line is the curve of heat; the dotted line, curve of luminosity.

mation, as I think, the true distribution of solar energy, as it reaches us after absorption by our own terrestrial atmosphere. There are some important corrections to be applied hereafter, such as those due to the selective absorption of the reflecting metals employed, which, as I have mentioned, though existing in less degree than with the prism, does yet exist. These will in some degree modify the form of the curve, which the reader is again invited to remember is given here only as a first approximation.

We can already, though, see here that there is nothing corresponding to the so-called "actinic" curve whatever. Precisely where this is represented at its maximum, in the ultra-violet, the real energy is nearly at its minimum. The special sensitiveness of certain salts of silver, then, to these radiations—not any special energy in the radiations themselves—led to the former quite mistaken belief that there was a something here called actinism, or chemical force, and to the belief, even still entertained (and also a mistake), that there is any considerable energy in this ultra-violet part,* to be interpreted by photography or in any other way.

In the fact that the whole energy perceptible by these means ceases at about wave-length ·00035 millimetres, while vision, without any special precautions, recognizes lines at ·0004 millimetres, we observe how very small the extent of the ultra-violet part of the spectrum really is. Photography can recognize a little more, but, of the whole of the energy in this portion, so much has been absorbed, either in the sun's atmosphere or our own, that what we get is insignificant.

We may further observe that the maximum of our curve falls in the orange or orange-yellow, not in the ultra-red. The sun's most energetic radiations, then, are *not* the invisible ones, as has been so long supposed, but the wave-length representing the maximum of heat does not differ very widely from that representing the maximum of light.

We may observe, nevertheless, that radiations of indefinitely great wave-length are found in the solar spectrum, for our own measures, extending far into the ultra-red, fail to reach any part where it suddenly ceases, as it does at the ultra-violet end. These are general conclusions which may apparently be safely drawn from

---

* In the ultra-violet, that is, *as we receive it.* It will appear probable, from what follows, that most of the sun's ultra-violet rays are absorbed in our upper atmosphere, and never reach us directly.

these approximate results; but, for further profitable study of the curve, we must wait for additional experiment and observation. There is, however, another and independent use to be made of observations of this kind, of great importance.

The amount of heat the sun sends the earth, or "the Solar Constant," as has already been observed in this work, has been measured by Herschel and Pouillet, and by numerous later observers. The most probable value assigned by the most recent and trustworthy investigators is about 24 calories, but this (as the text mentions) is not to be considered certain. Now, to see how the apparatus just described can be used to determine this solar constant, we have only to remember what has been said about the means of measuring the sun's heat, and then to consider that the extremely delicate strips of the Bolometer, which are heated by the sun to their utmost capacity in less than a second, constitute an instrument for determining the solar heat of the static class. They are assimilable in this aspect to the instrumental division in which M. Violle's actinometer comes, only that the Bolometer reaches its condition of equilibrium in a single moment, so to speak. As each ordinate of our curve represents the heat which is found at that point, the whole area between the curve and the horizontal line will represent the *whole* of the sun's heat which reaches us (save for the portion unmeasured, extending beyond the wave-length, ·0012 millimetres). If we could take our measuring apparatus *outside* our atmosphere, then, and repeat these observations, we should find a second curve where the ordinates (perpendiculars) were greater, and where the whole inclosed area was consequently greater also. As this second area represents the heat of the sun before absorption, it may be considered to represent the solar constant.

To determine the solar constant, then, by this method, we have to find what each perpendicular would be if it were drawn from measures made outside our atmosphere, and this again (impracticable as it may at first seem) is in reality quite easily ascertainable when we know the rate at which our atmosphere has absorbed each part.

It has already been mentioned in the text that there is a different rate of absorption for different parts of the spectrum; but what this is has never been exactly ascertained, because we have hitherto had no means of determining the energy in nearly homogeneous rays—those which fall on the thermometer, photometer, etc., in ordinary use, being evidently highly complex.

The narrow strips of the bolometer, then, constitute a static actinometer, to use which in finding the solar constant we measure ray by ray in the spectrum. If, for instance, the bolometer exposes a surface of one square centimetre, and if at noon, when the sun shines through a mass of air, which we will call 1, the energy of a certain ray in the ultra-violet is 20 on our (arbitrary) galvanometer scale, and if again later in the day, at some hour when the rays of the sinking sun pass through a mass of air represented by 2, our scale reads but 5 for the same ray, we find that as the energy in this ray after passing through *two* strata was to the energy after passing through *one* stratum, so is the energy after passing through that one stratum to the original energy before it entered the air. It is, in other words, a sum in rule of three, where, as five is to twenty, so is twenty to the answer, and the energy before absorption in the case of this instance is plainly 80, which is the amount of heat our instrument would have measured for this ray if transported outside our atmosphere.

For some other ray (for instance, one in the ultrared) we might have found quite a different rate of absorption. Thus: Supposing its energy measured at noon to have been 100, and again, when the absorbing mass of air was double, that it was 80, we see that this ray was far less absorbed than the other. The original ray in this case must plainly have had an energy of 125. So we can go on, rebuilding our perpendiculars to the height they must have to represent the energies before absorption; having done which, and having finally measured the energy represented by the inclosed curve, we find what total heat would have fallen on a surface one centimetre square outside our atmosphere in one second or one minute, and from this the heat in calories as compared with the same heat at the sea-level is instantly deducible. The result, which can not yet be given in detail, is, that the solar constant is *larger* than has been supposed, and probably much larger.

But this is not all, for evidently, when we have the different rates of absorption determined, our perpendiculars may grow in very different proportion, so that the *form* of the curve without our atmosphere may be quite different from that within.

It appears to be highly probable, from the observations thus far made, that the maximum ordinate in the extrá-atmospheric curve lies much nearer to the violet than it does in the curve after absorption, and that in fact the "center of gravity" of the curve as a whole is translated toward the violet, though how far toward it I am not prepared to here state. That is to say, that if the *eye* were outside our atmosphere, the totality of solar radiations would give it a sensation to which we should affix the word "blueness" rather than "yellowness" or "whiteness." Media of our atmosphere (and, it may be added, of the sun's atmosphere also), which we commonly think of as transparent, are then in reality playing a part analogous to that of a yellowish or reddish glass, whose impure color is not a monochromatic yellow or red, but a compound of many or even all the spectral tints in unaccustomed proportions. Had we in all our lives had no light but the electric light, seen only through such a reddish glass shade, we should doubtless believe this reddishness the "natural" color of the glowing, naked carbons, and the sum of all radiations. It would apparently answer (to a race brought up in ignorance of any other light) to our notion of *whiteness*. Its color would then seem to be no "color" at all, and the medium would, in this case questionless, be deemed transparent (as we believe our air transparent); and, if this medium were removed and the electric light seen in its true whiteness, it could not but seem that *it* was strongly colored.

Without pushing the analogy too far, then, the reader is invited to consider that, at any rate, these observations prove that neither he nor any one has ever seen the sun's face as it really is, and that it is at least not improbable, from what has been shown by the bolometric measures described above, that, if he could see it, he would pronounce it to be *blue*.

NOTE.—We add, as supplementary to Professor Langley's most interesting remarks, another diagram which shows clearly the apparent, striking divergence between his results and those obtained by nearly all pre-

APPENDIX. 307

vious observers, Dr. J. W. Draper alone excepted. Their results were obtained from the *prismatic* spectrum, instead of the diffraction spectrum, a fact which makes it difficult to compare them accurately with Professor Langley's, because, in the invisible spectrum, below the red, there are no data by which we can determine exactly the wave-lengths corresponding to points in the spectrum formed by the particular prisms used by them. By substituting for wave-lengths their reciprocals, we get, however, a scale for the spectrum which enough resembles that of an ordinary prism to allow the comparison to be established without gross error. The full line represents the heat-curve, as found by Professor Langley in the

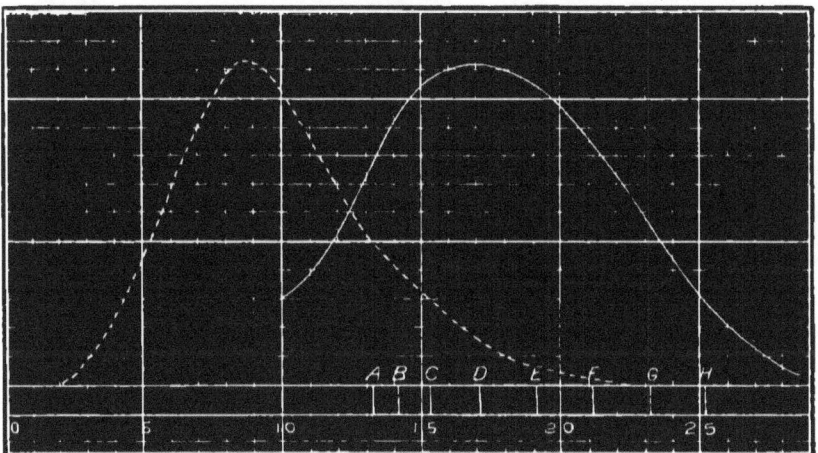

FIG. 62.

diffraction spectrum; the dotted line, the heat-curve given by Secchi as the result of the observations of Tyndall and others. We have called the divergence apparent, because it is in great part to be explained by the fact that, in the diffraction spectrum, the lower regions are enormously dispersed as compared with the upper. To compare results obtained in the prismatic spectrum with those from the *normal* or diffraction spectrum, it would therefore be necessary to multiply each result obtained in the former by the fraction expressing the ratio between the dispersions of the two spectra at the point in question, since the surface of the thermopile, or bolometer, by means of which the measure is made, has necessarily a considerable width—is not, and can not be, a mathematical line. We have not the necessary data as to the prisms used to enable us to make the correction now; but it is certain that, if made, it would tend greatly to reduce the discrepancy. It is very probable, also, that selective absorption by the glass of the prisms, and possibly also selective *reflection* by the metal of the diffraction grating and of the bolometer-strips, may have considerable influence. It is perfectly evident that the matter needs further careful study in order to determine the real amount of the discrepancy and its causes, and to reach the exact facts as to the distribution of heat in the spectra formed in different ways—a research upon which Professor Langley has already entered.

INDEX.

ABBE, extent of corona in eclipse of 1878, 222.
Absorption of sun's rays by the atmosphere of the earth, 262.
Actinic or chemical rays, 298.
Adjustment of focal plane of telescope to the slit of the spectroscope for observations upon the spectrum of the chromosphere, 195.
— — slit of spectroscope for observations of the prominences, 197.
Age and duration of the sun, 275–277.
Airy, solar parallax from the transit of 1874, 33.
Allotropic states of chemical elements, 90, 101.
American method of photographing the transit of Venus, 37.
Analyzing spectroscope, 77.
Andrews, critical temperature of a gas, 284.
Ångstrom, early studies in spectrum analysis, 67, 81.
— map of solar spectrum, 80, 90, 92, 93, 193, 230, 231, 233.
Animal, body of, regarded as a machine, 13, 14.

Arago, diminution of brightness at the limb of the sun, 245.
Aristarchus, method of determining the sun's parallax, 24.
Ascension Island, 29.
Aurora borealis, its spectrum not to be identified with that of the corona, 233.
— — relation to sun-spots, 156.
— — resemblance between its streamers and those of the corona, 238.
Axis of the sun, 138, 139.
— — — — table giving its position-angle for different times of the year, 139.

BASIC lines in solar spectrum, 91–93.
Barker, dark lines in spectrum of the corona, 234.
Belli, photometric observation upon the brightness of the corona, 228.
Bessemer converter, compared with the solar radiation, 245, 268, 279.
Biela, brightness of the inner corona, 229.
Blueness of sunlight before suffering atmospheric absorption, 250, 306.

INDEX.

Bolometer described, 300.
— determination of true value of the solar constant, 304.
— sensitiveness of, 300.
Bond, method of determining the sun's parallax by observations of Mars, 25.
Bouguer, measurement of the sun's light, 241.
Brightness of the corona, 228–230.
Bullock, drawing of eclipse of 1868, 220.
Bunsen, arrangement of spectroscope scale, 70.
— work upon the solar spectrum in connection with Kirchhoff, 67.
Burning-glass, effect of, 268.

CANDLE-POWER, or photometric unit, defined, 241.
Calory, or thermal unit, defined, 255.
Calcium-light compared with sunlight, 244. 279.
Capocci, theory that spots are due to volcanic eruptions on the sun, 167.
Carrington, discovery of sun's equatorial acceleration and formula for it, 133, 134.
— distribution of sun-spots, 200.
— method of determining the position of a spot on the sun, 52.
— motion of spots in latitude, 140.
— observation of remarkable solar outburst, November 1, 1859, 119, 156.
— Period of sun's rotation, 133.
— Position of sun's axis, 138.
Cassini, observations for solar parallax, 28.
Chambers, barometric effect of sun-spots, 162.

Chapman, ruling of diffraction gratings, 73.
Christie, solar eyepiece, 65.
Chromatosphere, or chromosphere, defined, 17, 180.
Coal, consumption of which would be required to keep up the solar radiation, 255.
Comets' tails, their analogies to the streamers of the corona, 238, 293.
Comparison-prism, 85.
Condensation theory of solar heat, 273–275.
Constancy of solar heat during the historic period, 270.
Constitution of sun, 18, 280–290.
Contact observations at the transit of Venus, 32, 33.
— — by means of photography, 35.
Corona, brightness of, 228–230.
— defined, 17.
— examined by slitless spectroscope, 234, 235.
Corona-line in the spectrum, discovery, 224.
— — duplicity of, 230.
— — map, 231.
— — not identical with line in spectrum of aurora borealis, 233.
Cornu, determination of the velocity of light, 42.
— solar photography, 54.
Critical temperature of a gas, 284.
Croll, hypothesis that a portion of the sun's energy may have originated in a collision with a star, 277.
Crova, pyrheliometer, 258.
— value of solar constant, 263.
Cyclonic motion in sun-spots, 124, 172, 173.

INDEX. 311

DALTON, law of gaseous mixture, 287, 288.
Dark lines in the solar spectrum discovered, 66.
— — explanation of, 81, 82.
— — in spectrum of the corona, 234.
Davis, photograph of eclipse of 1871, 224.
Dawes, "holes" in nucleus of sun-spot, 117.
— solar eyepiece, 65.
De La Rue, the Kew photoheliograph, 55, 56.
— — — photographs of the eclipse of 1860, 183.
— — — measures of sun-spot penumbra, 127.
— — — planetary influence on sun-spot development, 149.
— — — relation of Wolf's "relative numbers" to the spotted area of the sun, 147.
Denza, bright lines in corona spectrum, 233.
Derham, volcanic theory of sun-spots, 167.
Detached cloud-formed prominences and their development, 206.
Development of sun-spots, 120, 121.
Deville, estimate of the temperature of the sun, 265.
Diameter and dimensions of the sun, 45, 278.
— — illustrations, 46, 281.
Diffraction grating, 73.
— spectroscope, 74, 75.
— spectrum, 76.
Dimensions of sun-spots, 125.
Diminution of brightness at limb of the sun, 52, 108, 245-250, 279, 292.
Discovery of bright line in corona spectrum, 224.

Discovery of dark lines in solar spectrum, 66.
— — dark lines in spectrum of corona, 234.
— — elements present in the sun, 87, 88.
— — equatorial acceleration of the sun, 133.
— — explanation of cause of dark lines, 81.
— — gaseous constitution of the prominences, 185.
— — magnetic relations of sun-spots, 154.
— — oxygen in the sun, 94.
— — periodicity of sun-spots, 144.
— — reversing layer of the sun, 83.
— — spectroscopic method of observing prominences, 186.
— — sun-spots, 113.
Displacement and distortion of lines by motion, 97-100, 195.
Dissolution and disappearance of sun-spots, 121, 122.
Distances (relative) of planets, 26.
Distance of the sun from the earth, illustrations, 43-45.
Distortion of forms of prominences by spectroscope, 192.
Distribution of sun-spots and prominences in solar latitude, 140-142, 200.
— — energy in solar spectrum, 299, 301, 307.
Don Ulloa, observation of "hole in the moon," in the eclipse of 1778, 182.
Draper, Dr. Henry, discovery of oxygen in the sun, 94-96.
Draper, J. W., early spectroscopic researches, 67.
— — distribution of energy in solar spectrum, 307.

Draper, John C., dark lines of oxygen, 96.
Drawings of corona, discrepancies, 215.
Dulong and Petit, law of specific heats, 92.
— — — — radiation, 266.
Duration of sun-spots, 118.

EARTH, dimensions of the, 22.
— her share of the solar radiation, 256.
Eastman, photometric observations during eclipse of 1869, 227.
Eclipse, solar, 1706—182; 1715—182; 1733—181; 1778—182; 1806—82; 1842—183, 228; 1851—183; 1857—216; 1860—183, 217, 218, 229; 1867—219; 1868—184, 185, 220; 1869—221, 224, 227, 233; 1870—84, 225; 1871—222-225, 229, 235, 238; 1878—215, 225, 237.
— — general phenomena, 213, 214.
Ecliptic, defined, 16.
Effect of changes in solar atmosphere upon terrestrial conditions, 264.
Effective temperature of the sun, 266.
Electric light compared with sunlight, 241, 279.
Elements known to be present in the sun, table, 87, 88.
Encke, discussion of transits of Venus in 1761 and 1769, 31.
Energy (total) of solar radiation, 255.
— — — distribution in the spectrum, 299, 301, 307.
Energy, terrestrial, mainly derived from solar heat, 12, 13.

Energy, terrestrial—other sources than solar heat, 14.
Equatorial acceleration of the sun, 133-138.
— — explanation suggested, 136, 137.
Ericsson, estimate of the sun's temperature, 265.
— experiment upon radiation of molten iron, 268.
— measure of solar heat, 259.
— solar engine, 256.
Eruptive prominences, 207, 208.
Experiment, showing that the blackness of the dark lines in the spectrum is only relative, 82.
Explanation of the sun's eruptive action caused by the constriction of the photosphere, 212.

FABRICIUS, discovery of sun-spots, 113.
Faculæ, 106-108.
Faye, explanation of the sun's equatorial acceleration, 138.
— formula for the acceleration, 135.
— theories of sun-spots, 170, 172-174.
— computation of solar parallax, 31.
Ferrers, observation of eclipse of 1806, 182.
Fizeau, comparison of electric and calcium light with sunlight, 244.
Flamsteed, method of deducing solar parallax from observations of Mars, 25.
Flashes seen by Peters in sun-spots, 123.
Fœnander, drawing of eclipse of 1871, 223.

INDEX. 313

Forbes, value of solar constant, 263.
Foucault, comparison of electric and calcium lights with sunlight, 244.
— determination of the velocity of light, 42.
— early spectroscopic researches, 67.
Fourteen hundred and seventy-four line, 230–235.
Frankland names *helium*, 88.
Fraunhofer, discovery of dark lines in solar spectrum, 66.
— coincidence of D line in solar spectrum with bright line in flame spectrum, 81.
— map of the spectrum, 79.

GALILEO, discovery of sun-spots, 113.
— theory of sun-spots, 167.
Gas, Dalton's law of mixture, 287, 288.
— distinctive properties, 287, 288.
— Lane's law of temperature and condensation, 274.
— liquefaction and critical temperature, 284.
— viscosity at high temperatures, 286.
Gaseous condition of the sun's nucleus, 282–286.
Gautier, relation between magnetism and sun-spots, 154.
Gill, observations of Mars for determination of solar parallax, 29, 30.
Gilliss, observations in Chili for the solar parallax, 29.
Gilman, corona of eclipse of 1869, 220.
Gitter (see Grating).

Gould, diminution of the earth's temperature at sun-spot maximum, 161.
Granulation of the sun's surface, 102–105.
Grant, early recognition of the chromosphere, 183, 188.
Grating, diffraction, used in spectroscope, 73–75, 93, 192.
Greenwich magnetic record for August 3 and 5, 1872, 158.
Gregory first calls attention to transits of Venus as a means of determining the sun's parallax, 31.
Grosch, drawing of eclipse of 1867, 219.

H LINES reversed in the spectrum of sun-spots, 131.
Habitability of the sun, 168, 285.
Halley, determination of the sun's parallax by transit of Venus, 31.
Hansen, detection of error in the received value of the sun's parallax, 31, 39.
Harkness, discovery of the bright line in the corona spectrum, 224.
Hastings, comparison of the spectrum of the sun's limb with that of the central portion of the disk, 84.
— smoke-like nature of the layer which causes the darkening of the sun's limb, 248, 292.
— theory of the constitution of the sun, 290–295.
Heat-curve of solar radiation as determined by the thermopile and bolometer, 301, 307.
Heat derived from stars and meteors, 14.
Heliometer described, 29.

INDEX.

Helioscopes and helioscopic eye-pieces, 61–64.
Helium and its characteristic line, 88, 233.
Helmholtz, condensation theory of the solar heat, 274.
Henry, observations with the thermopile upon radiation of sun-spots and different portions of the sun's disk, 159, 263.
Herschel, Sir John, measurement of the sun's heat, 252–255.
—— —— meteors as the cause of the sun's equatorial acceleration, 135.
—— —— solar eyepiece, 62.
—— —— theory of sun-spots, 169.
—— —— use of a prism in connection with diffraction grating to separate spectra of different orders, 75.
Herschel, Captain John, observation of prominence spectrum in 1868, 185.
Herschel, Sir W., relation between sun-spots and price of wheat, 145.
— theory of the sun-spots and the sun's constitution, 168.
Hodgson, observation of solar outburst in 1859, 119.
Horrebow, anticipation of the periodicity of sun-spots, 145.
Huggins, granulation of the sun's surface, 105.
— use of widened slit in observing forms of prominences, 189.
Hydrogen-lines in the spectrum of the corona, 231, 232.

ICE, quantity which would be melted in a minute by the sun's radiation, 254, 255.

Intra-Mercurial planet, 53, 273.
Investigation as to the influence of the planets upon the generation of sun-spots, 149, 150.

JANSSEN, discovery of method of observing prominences by means of the spectroscope, 185, 186.
— medal from French Government, 188.
— observation of the eclipse of 1868, 186.
— observations of the eclipse of 1871 and recognition of bright lines of hydrogen and dark Fraunhofer lines in the corona spectrum, 232, 234.
— observation of Venus on the corona, 229.
— photographic contact at the transit of Venus, 35.
— *Reseau Photospherique*, 110–112.
— solar photography, 59, 110.
Jelinek, influence of sun-spots on the temperature of the earth, 161.
Jevons, connection between sun-spots and commercial crises, 165.
Jupiter, influence upon sun-spots, 149, 150.

KEW, photoheliograph, and photographic record, 55–58.
Kirchhoff, map of solar spectrum, 130, 210, 230.
— spectroscopic work, 67, 81, 82, 87.
— theory of sun-spots, 167.

LACAILLE, observations for solar parallax, 28.
Lalande, theory of sun-spots, 167.
Lambert, diminution of light at the limb of the sun, 245.

Lane, estimate of the sun's temperature, 265.
— law relating to the temperature of a contracting mass of gas, 274.
Langley, bolometer and bolometric observations, 298–307.
— color of the sun's limb compared with that of the center of the disk, 248.
— comparison between the intensity of solar radiation and that of the metal in a Bessemer converter, 245, 268.
— details of the solar surface (*frontispiece*), 103.
— diminution of heat at the sun's limb, 263.
— diminution of light at the sun's limb, 247.
— effect of the sun's atmosphere and its changes upon the earth's temperature, 264.
— extent of corona in eclipse of 1878, 222.
— increase of solar radiation due to disturbance of the sun's surface, 160.
— observation of Mercury at the transit in 1878, 229.
— solar eyepiece, 65.
— spectroscopic observation of the sun's rotation, 100.
— temperature of sun-spots, 159.
— thermopile observations, 263, 264.
— true color of the sun, 251, 306.
Laplace, effect of the absorption of the atmosphere of the sun upon its brightness, 249.
Laugier, sun's equatorial acceleration, 133.

Laussedat, horizontal photoheliograph, 36.
Lens, burning effect of, 268.
Leverrier, determination of the parallax of the sun by means of planetary perturbations, 26.
— perturbations of Mercury indicating intra-Mercurial planets, 273.
Liais, drawing of eclipse of 1857, 216.
Light of the sun, its total quantity in standard candle-power, 240, 279.
— — — — its intensity, 244, 279.
— — — — method of measuring, 242, 243.
— velocity of, used in determining the solar parallax, 26, 41, 42.
Lindsay, Lord, expedition to Ascension Island, 29.
Liquefaction of gases, 284.
Lockyer, arrangement for studying the solar spectrum, 84.
— connection between sun-spots and rainfall in India and Africa, 164.
— discovery of the spectroscopic method of observing the chromosphere and prominences, 186–188.
— discovery of the 1474 line in the *chromosphere* spectrum, 230.
— medal from the French Government, 187.
— observation of the lines of hydrogen in the corona spectrum, 232.
— theory as to the non-elementary character of so-called elements, 89–94.
— use of annular slit for observing circumference of the sun, 199.
— vibrating slit for observation of prominences, 188.

Loomis, effect of conjunctions of Jupiter and Saturn upon sun-spot periodicity, 150.
— relation between sun-spots and the aurora borealis, 156.
Luminous radiations, falsely distinguished from thermal and chemical, 298.
Lunar perturbations, as a means of determining the solar parallax, 39.

MAGNETISM, terrestrial, period of disturbance corresponding with the sun-spot period, 153-155.
— — affected by solar paroxysms, 119, 120, 156-158.
Mars, observed as a means of determining solar parallax, 27, 28.
— opposition of 1877, 29.
Mass of the sun, 46, 278.
Maxwell, effect of temperature upon the viscosity of a gas, 286.
Mechanical equivalent of heat, 271.
Medal struck by the French Government in honor of Janssen and Lockyer, 187.
Meldrum, connection between sun-spots, cyclones, and rainfall, 162-164.
Mercury (planet), influence on sun-spots, 149.
— perturbations indicating intra-Mercurial matter, 273.
— seen at transit on the background of the corona, 229.
Merz helioscope, 63.
Metallic prominences, 209.
Metals, present in the sun, 87, 88.
Meteors, possibly concerned in the formation of sun-spots, 151.

Meteors, regarded as the cause of the sun's equatorial acceleration, 135.
Meteoric theory of the sun's heat, 271, 272.
Meudon, solar observatory, 59, 111, 166.
Michelson, determination of the velocity of light, 42.
Mouchot, solar engine, 256.

NASMYTH, willow-leaf structure of the sun's surface, 104.
Newcomb, determination of the solar parallax, 43.
— extent of the corona in the eclipse of 1878, 222.
— reduction of Gilliss's observations at Santiago, 29.
— speculations as to the age and duration of the sun, 276.
Nodes of the sun's equator, 139.

OXYGEN in the sun, Dr. H. Draper, 94, 95.
— spectra of, Schuster, 96.

PARALLAX, solar, defined, 22.
— — determined by lunar perturbations, 39, 40.
— — determined by observations of Mars, 27-30.
— — determined by planetary perturbations, 40, 41.
— — determined by transits of Venus, 30-39.
— — determined by the velocity of light, 42.
— — importance and difficulty of the problem, 20-23.
— — synopsis of methods for its determination, 25, 26.
— — values, according to different authorities, 42, 43, 278.

INDEX.

Peters, observations of sun-spots, 122, 123.
— volcanic theory of sun-spots, 168.
Petit, observation of the corona in 1860, 229.
Photographic observations of transit of Venus, 34–39.
Photographs of eclipses — 1860, 183; 1871, 217, 224.
Photography, solar, 54–60, 110–112.
Photometric observations upon the corona, 226–229.
Photosphere defined, 17.
— theories as to its nature, 109, 138, 171, 175, 177, 289–291.
Picard, observations for solar parallax, 28.
Pickering, diminution of light at the sun's limb, 246, 249.
— effect of the sun's atmosphere upon its brightness, 250.
— solar eyepiece, 65.
Pitch of a sound altered by motion, 98.
Planetary perturbations as a means of determining solar parallax, 40.
— influence upon sun-spots, 149–151.
Planets, determination of their relative distances, 26.
Pogson, observation of eclipse of 1868, 185.
Polarization of the corona, 234.
Polarizing eyepieces or helioscopes, 64, 65.
Position-angle of the sun's axis — table, 139.
Potsdam, astrophysical observatory, 166.
Pouillet, estimate of the sun's temperature, 265.
Pouillet, measurement of the sun's heat, 252, 257, 263.
— pyrheliometer, 257.
— temperature of space, 14.
Powalky, computation of solar parallax, 31.
Princeton, Henry's thermopile observations, 159, 263.
— spectroscope used in the observatory, 75.
Prisms and prismatic spectrum, 68–72.
Problems in solar physics, 297.
Proctor, demonstration that the corona can not be due to the earth's atmosphere, 226.
— discussion of Jevons's paper on connection of sun-spots with financial crises, 165.
— velocity of matter ejected from the sun, 212.
Projection of the sun's image on a screen, 50.
Prominences or protuberances (solar) defined, 17.
— — — first named, 180.
Purple tint of the nucleus of a sun-spot, 117.
Pyrheliometer, 257, 258.

QUIESCENT prominences, 204.

RADIATION (total) of the sun, 255, 256, 279.
Ranyard, brightness of the inner corona, 229.
— memoir on recent eclipses, 216.
— synclinal structure of the corona, 237.
Rayet, observation of the eclipse of 1868, 185.

Rayleigh, Lord, resolving power of spectroscopes, 73.
Recurrence of sun-spots at special points on the sun's surface, 143, 289.
Reflecting telescope with unsilvered mirror for observing the sun, 61.
Reseau Photospherique, Janssen, 110–112.
Respighi, depression of the chromosphere over a sun-spot, 202.
— observation of the corona with a slitless spectroscope, 235.
Reversal of bright lines to dark in the solar spectrum explained, 82.
— of dark lines to bright at a total eclipse, 83.
— double, of D lines in the chromosphere spectrum, 196.
Reversing stratum of the solar atmosphere first observed, 83, 84.
— — its relation to the photosphere, 280, 294.
Richer, observations for solar parallax, 28.
Roemer, observations for solar parallax, 28.
"Rosa Ursina," Scheiner, 114.
Rosetti, law of radiation and effective temperature of the sun, 267.
Ross, photometric observations upon the corona, 228.
Rotation of the sun demonstrated by displacement of lines in the spectrum, 100.
— — — — peculiar law of equatorial acceleration, 133–138.
Rutherfurd, diffraction gratings, 73.
— solar photography, 54.

SANTIAGO, observations of Gilliss, 29.

Saturn, influence on sun-spots, 150.
Scheiner, discovery of sun-spots, 113.
Schott, drawing of eclipse of 1869, 221.
Schuster, spectra of oxygen, 90.
Schwabe, discovery of the periodicity of sun-spots, 144.
Secchi, classification of prominences, 204.
— drawing of eclipse of 1860, 217.
— drawing of a sun-spot, 115.
— estimate of the sun's temperature, 265.
— formation of detached cloud-like prominences, 206.
— measurement of the variations of temperature at different parts of the sun's disk, 263, 264.
— photographs of the eclipse of 1860, and inferences from them, 183, 184.
— solar eyepiece, 65.
— thermopile observations, 263, 264.
— theories of sun-spots, 170–174.
Sherman, observations at, 157, 192, 210.
Sierra, synonym for chromosphere, 180.
Silvered object-glass for viewing the sun, 61.
Slitless spectroscope applied to the corona, 234, 235.
Smyth, records of rock-thermometers at Edinburgh, 162.
Solar, constant, defined, 263.
— — value of, 263, 279, 303–305.
Solar parallax (see Parallax).
Soret, penetrating power of solar radiation, 269.
Sources of solar heat, 270–274.

Space, temperature of, 14.
Spectral photometer, Vogel, 246.
Spectra produced by prisms and gratings compared, 76.
Spectroscope, analyzing and integrating, 76, 77.
— automatic, 190.
— described and discussed, 67–77.
Spectrum, explanation of its formation in a spectroscope, 69.
— of the corona, 230, 231.
— — a sun-spot, 129.
— solar, discovery of the dark lines, 66.
— — early investigations as to the origin of the dark lines, 67.
— — Kirchhoff's explanation of the dark lines, 82.
— — maps or drawings of portions, 79, 80, 99, 129, 130, 157, 196, 210, 231.
Spoerer, distribution of sun-spots, 142.
— estimate of the sun's temperature, 265.
— formula for sun's equatorial acceleration, 135.
— recurrence of spots at special points on the sun's surface, 143, 289.
Spots (see Sun-spots).
Stannyan, Captain, discovers the chromosphere in 1706, 182.
Stewart, Balfour, area of sun-spots, 159.
— — discussion of magnetic observations at Kew, 147, 155.
— — uncertainty whether sun-spots raise or lower terrestrial temperature, 161.
Stone, calculation of solar parallax, 31–33.

Struve, brightness of the corona, 229.
Sun-spots, cyclonic motion of, 124.
— — depressions in the photosphere, 126, 128.
— — development and dissolution, 121, 123.
— — dimensions, 125.
— — discovery in 1610, 113.
— — distribution on the sun, 140, 142.
— — disturbances connected with them, 119, 192.
— — duration, 118.
— — effects upon the earth, 153–166.
— — periodicity, 144–152.
— — spectrum, 129.
— — theories as to formation and nature of, 166–177.
— — visible to the naked eye, 113, 125, 126.
Swan, spectroscopic observations, 67, 81.
Symons, connection between sun-spots and rainfall, 164.
Synclinal structure of the corona, 237.

TARDÉ, *sidera Borbonica*, 114.
Telespectroscope, 78.
Tempel, drawing of eclipse of 1860, 218.
Temperature of the sun, 265, 269, 279.
— — the sun's center, 286.
— — sun-spot, 159.
— terrestrial as affected by sun-spots, 160–162.
Tennant, Colonel, observation of eclipse of 1868, 185.
Thalen, elements represented in the solar spectrum, 93.

Thickening and thinning of lines in the sun-spot spectrum, 129.
Thermal rays falsely distinguished from luminous and chemical, 298.
Thermopile, 263, 300.
Thollon, powerful spectroscopes, 72, 92.
Thomson, Sir W., endurance of the sun's heat if produced by the combustion of coal, 256.
— — — estimate of heat which would be produced by the fall of planets on the sun, 272.
Tisserand, formula for sun's equatorial acceleration, 135.
Todd, value of solar parallax deduced from the velocity of light, 42.
Transit of Venus, 30–40.
Trouvelot, veiled spots, 132.
Tupman, drawing of the eclipse of 1871, 222.
— work upon solar parallax, 33, 39.
Tyndall, distribution of heat in the solar spectrum, 307.

ULLOA, Don, observation of eclipse of 1778, 182.

VARIATIONS in solar radiation, 265.
Vassenius, early observation of prominences, 181.
Veiled sun-spots, Trouvelot, 132.
Velocity of motion in solar prominences, 209, 210.
Venus, influence upon sun-spots, 149.
— seen at transit before reaching the limb of the sun, 229.
Vicaire, estimate of the sun's temperature, 265.
Violle, actinometer, 260.

Violle, measure of the sun's heat, 259–261.
— value of the solar constant, 263.
Vogel, diminution of light at the sun's limb, 246, 247, 250.
— effect of the sun's absorbing atmosphere upon his total brightness, 250.
— exposure slide for solar photography, 57.
— spectral photometer, 246.
— spectroscopic measurement of the sun's rotation, 100.

WARRING, illustrations of the sun's attracting force on the earth, 48.
Waterston, measure of solar heat, 259.
Wilna, photographic observations, 59.
Wilson, discovery that sun-spots are depressions in the sun's surface, 126.
Winlock, horizontal photoheliograph, 36.
— annular slit for spectroscopic observation of the prominences, 199.
Wolf, magnetic variations following sun-spot period, 155.
— periodicity of sun-spots and relative numbers, 145–147.
Wollaston, discovery of dark bands in the solar spectrum, 66.
— measurement of the sun's light, 241.

YOUNG, discovery of bright lines in the spectrum of the corona, 224, 233.
— disturbance of lines in sun-spot spectrum, 99, 130.

Young, double reversal of D lines, 196.
— duplicity of corona line, 230.
— examination of basic lines in the solar spectrum, 92, 93.
— experiment showing the blackness of dark lines to be only relative, 82.
— observations on chromosphere lines at Sherman, 192.
— observations on the corona at Denver in 1878, 215.
— observations on remarkable prominences, 202, 206-208.
— proposed explanation of equatorial acceleration, 136, 137.
— reversal of dark lines at the beginning of totality in the eclipse of 1870, reversing the stratum, 83.

Young, solar eruption followed by magnetic disturbance, 157, 158, 210.
— spectroscopic measurement of the sun's rotation, 100.
— sun-spot spectrum, 129, 130.

ZANTEDESCHI, development of the spectroscope, 67.
Zöllner, estimate of the sun's temperature, 265.
— formula for the sun's equatorial acceleration, 135.
— spectroscopic measurement of the sun's rotation, 100.
— theory of sun-spots and liquid surface of the photosphere, 171.
— vibrating slit for observation of the prominences, 188.

THE END.

www.ingramcontent.com/pod-product-compliance
Lightning Source LLC
Chambersburg PA
CBHW030750230426
43667CB00007B/916